T0215594

Building Enterprise IoT Solutions with Eclipse IoT Technologies

An Open Source Approach to Edge Computing

Frédéric Desbiens

APress®

Building Enterprise IoT Solutions with Eclipse IoT Technologies: An Open Source Approach to Edge Computing

Frédéric Desbiens
Embrun, ON, Canada

ISBN-13 (pbk): 978-1-4842-8881-8   ISBN-13 (electronic): 978-1-4842-8882-5
https://doi.org/10.1007/978-1-4842-8882-5

Managing Director, Apress Media LLC: Welmoed Spahr
Acquisitions Editor: Jonathan Gennick
Development Editor: Laura Berendson
Coordinating Editor: Jill Balzano

Cover photo by Szabo Viktor on Unsplash

Distributed to the book trade worldwide by Springer Science+Business Media LLC, 1 New York Plaza, Suite 4600, New York, NY 10004. Phone 1-800-SPRINGER, fax (201) 348-4505, e-mail orders-ny@springer-sbm. com, or visit www.springeronline.com. Apress Media, LLC is a California LLC and the sole member (owner) is Springer Science + Business Media Finance Inc (SSBM Finance Inc). SSBM Finance Inc is a **Delaware** corporation.

For information on translations, please e-mail booktranslations@springernature.com; for reprint, paperback, or audio rights, please e-mail bookpermissions@springernature.com.

Apress titles may be purchased in bulk for academic, corporate, or promotional use. eBook versions and licenses are also available for most titles. For more information, reference our Print and eBook Bulk Sales web page at http://www.apress.com/bulk-sales.

Any source code or other supplementary material referenced by the author in this book is available to readers on GitHub.

Printed on acid-free paper

Avec tout ce que je sais, on pourrait faire un livre...
il est vrai qu'avec tout ce que je ne sais pas,
on pourrait faire une bibliothèque.

With everything I know, I could create a book...
and with everything I know not, I could create a library.

−Sacha Guitry, Le KWTZ, in Pièces en un acte

This book is dedicated to the contributors, committers, and project leads of the Eclipse Foundation. Without your dedication and commitment to open source, this book would not exist.

Table of Contents

About the Author xv

About the Technical Reviewers xvii

Acknowledgments xxi

Introduction xxiii

Part 1: Fundamentals and Protocols 1

Chapter 1: What Is IoT? 3

    IoT and Other Current Trends 4

    Edge Computing and IoT 6

    IoT Reference Architecture 9

        Common Layers 10

        Constrained Devices 12

        Edge 14

        IoT Platforms 16

    Protocols: Foundational Building Blocks 20

    An Extensive IoT and Edge Toolkit 22

Chapter 2: CoAP 25

    CoAP: A Slimmer HTTP 26

    Characteristics 26

    Protocol Stack 29

    Security 31

    Eclipse Californium 32

        Sandbox Server 33

        Getting Started 34

        Simple GET Request: Demo 34

Simple GET Request: The Code 35

Simple Server: Demo 37

Simple Server: The Code 38

About DTLS 40

CoAP and Constrained Devices 42

Chapter 3: LwM2M 45

LwM2M: Built on CoAP's Foundation 45

Data Model: Objects and Resources 46

Additional Capabilities over CoAP 48

Bootstrap 49

Client Registration 49

Device Management and Service Enablement 50

Information Reporting 50

LwM2M Versions 50

LwM2M v1.0 (February 2017) 50

LwM2M v1.1 (June 2018) 51

LwM2M v1.2 (November 2020) 51

LwM2M Protocol Stack 52

Eclipse Leshan 53

Sandbox Server 54

Quick Test Drive 54

Building Your Client 58

Building Your Own Server 61

LwM2M and Constrained Devices 63

Working with 6LoWPAN 65

Chapter 4: MQTT 67

What Is MQTT? 68

Messages 68

Topics and Topic Filters 70

Quality of Service 73

Publish and Subscribe 74

Connections and Sessions 76

Retained Messages 78

Protocol Stack 79

MQTT-SN 80

New Features in MQTT v5 82

    User-Defined Properties 83

    Reason Codes 83

    Availability of Optional Features 84

    Message Expiry 84

    Request/Response 84

    Clean Session 84

    Server DISCONNECT 85

    AUTH Packet 85

    Passwords Without Usernames 85

    Shared Subscriptions 85

    Flow Control 86

    LWT Delay 86

    Broker Reference 87

Security 87

Broker: Eclipse Mosquitto 88

    Installation 88

    Configuration 89

    Setting Up TLS 92

    Sandbox Servers 95

Building Clients with Eclipse Paho 96

    Java 96

    Python 98

MQTT and Constrained Devices 100

    Eclipse Paho Embedded Clients 100

    Zephyr RTOS 101

Chapter 5: Sparkplug 103

Sparkplug and Interoperability 104

Architecture 106

Sparkplug MQTT Requirements 107

Basic Principles 107

Topic Namespace 108

    Namespace 109

    Group ID 109

    Message Type 110

    edge_node_id 110

    device_id 110

Message Types 110

Payload Definition 112

    Metrics 113

Session Management 114

Example 114

    Host Application 115

    Edge Nodes 115

    Devices 115

Leveraging Eclipse Tahu 117

Chapter 6: DDS 121

What Is DDS? 122

Protocol Stack 123

Publish and Subscribe 125

Global Data Space 128

    Domains 128

    Partitions 129

Quality of Service 130

    Data Availability 130

    Data Delivery 131

Data Timeliness 133

Resources 133

Configuration 134

Topics 134

Topic Types 134

Topic Keys, Instances, and Samples.. 137

Filtering 137

Reading and Writing Data 138

Writing 139

Reading 140

Security 141

Eclipse Cyclone DDS 143

Installation 143

The Hello World! Sample 147

Cyclone DDS and Robotics 153

Chapter 7: zenoh.. 155

zenoh Basics 156

Key Abstractions 157

Primitives.. 159

Deployment Units 160

Device Discovery 162

Example: Robots at the Factory 163

Plugins 165

Storages Plugin Deep Dive 166

Reliability and Congestion Control 167

zenoh-flow 168

zenoh Protocol Stack 170

Coding with Eclipse zenoh 171

Installation.. 171

Scouting 178

Subscribing 179

Publishing 180

zenoh on Constrained Devices 181

    Installation 182

    Zephyr Development Environment 182

    Subscribing 183

    Publishing 184

Part 2: Constrained Devices **187**

Chapter 8: The Hardware **189**

    Selection Criteria 189

        Power Consumption 190

        Life Cycle 190

        Use Case Requirements 191

        Security 192

        Cost 193

    Microcontrollers 194

        Arm: Cortex-M 195

        AVR: The Heart of Arduino 198

        RISC-V: A Serious Challenger 199

    Choosing an MCU 201

    Sensors and Actuators 202

        Sensor Characteristics 202

        Making the Connection 204

    Endpoints, Gateways, and Edge Nodes 206

        Arm Cortex-A 207

        x86-64 211

    Open Source Hardware 212

        Processor Cores: OpenHW CORE-V Family 213

        Microcontroller: CORE-V MCU 214

        Single-Board Computers 214

Chapter 9: Connectivity **217**

  Connectivity Options 218

    Bluetooth 219

    Cellular (LTE, 5G) 221

    Ethernet 223

    Narrowband 224

    Wi-Fi 228

    Z-Wave 229

    Zigbee 230

  Network Design Considerations 232

    Choice of Protocols 232

    Confidentiality 232

    Resiliency 234

    Redundant Connectivity 235

    Out-of-Band Management 236

  Connectivity and Constrained Devices 236

    Wi-Fi 237

    Bluetooth 238

Chapter 10: Operating Systems **243**

  Are Operating Systems Necessary? 244

  Real-Time Operating Systems 246

    Arm Mbed OS 246

    FreeRTOS 250

    Zephyr 255

  Bare Metal Options 258

    Arduino 258

    Drogue IoT 259

    Espressif IDF (ESP-IDF) 261

Edge Nodes: The Realm of Linux 262

Getting Started with Zephyr 263

    Linux 263

    macOS 265

    Windows 266

Part 3: Edge Computing and IoT Platforms 269

Chapter 11: Edge Computing 271

What Is Edge Computing? 272

    Edge vs. Cloud 273

    The Edge-to-Cloud Continuum 273

    What Makes the Edge Different? 275

    What Are Edge Native Applications? 277

The Need for EdgeOps 278

Edge Computing Platforms 279

    Eclipse ioFog 281

    Eclipse Kanto 285

    Eclipse Kura 288

    Kubernetes 292

    Project EVE 294

Chapter 12: Applications 297

Application Runtimes 298

    .NET 298

    Java SE and Jakarta EE 299

    Node.js 300

    Python 302

    Rust 303

    WebAssembly 304

    How to Choose Your Runtime 304

Web of Things 305

    Eclipse EdiTDor.. 306

    Eclipse Thingweb... 308

Digital Twins... 310

    Eclipse Ditto... 311

How It All Fits Together 317

A Few Examples of Applications 319

    Eclipse 4diac 319

    Eclipse Keyple 321

    Eclipse VOLTTRON.. 322

Chapter 13: Integration and Data **325**

Device Connectivity... 326

    Eclipse Hono.. 326

    Getting Started with Hono 328

Device Management 333

    Eclipse hawkBit.. 334

    Eclipse Kapua 338

Data.. 341

    Eclipse Streamsheets 342

Chapter 14: Conclusion.. **347**

Index.. **351**

About the Author

Frédéric Desbiens manages IoT and edge computing programs at the Eclipse Foundation. His job is to help the community innovate by bringing devices and software together. He is a strong supporter of open source. In the past, he worked as a product manager, solutions architect, and developer for companies as diverse as Pivotal, Cisco, and Oracle. Frédéric holds an MBA in electronic commerce, a BASc in computer science, and a BEd, all from Université Laval. After hours, Frédéric will typically read a history book, play video games, or watch anime.

About the Technical Reviewers

Robert Andres is the Chief Strategy Officer at Eurotech Group, where he is responsible for driving edge computing, M2M, and IIoT initiatives within a wide range of markets and applications. Robert is an active contributor to the Eclipse IoT and Edge Native Working Groups at the Eclipse Foundation.

Simon Bernard is a Software Engineer at Sierra Wireless. He is the project lead for the Eclipse Leshan LwM2M implementation and the Eclipse Wakaama LwM2M client. He is also a committer on Eclipse Californium and Eclipse tinydtls.

Erik Boasson is Head of Technology at ZettaScale Technology. He is the project lead for Eclipse Cyclone DDS.

Angelo Corsaro is Chief Executive Officer (CEO) and Chief Technology Officer (CTO) at ZettaScale Technology. At ZettaScale, he is working with a world-class team to change the landscape of distributed computing – working hard to bring to every connected human and machine the unconstrained freedom to communicate, compute, and store – anywhere, at any scale, efficiently and securely. Angelo is a world top expert in edge/fog computing and a well-known researcher in the area of high-performance and large-scale distributed systems. Angelo has over 100 publications on referred journals, conferences, workshops, and magazines. Angelo has coauthored over ten international standards.

Ian Craggs has been working with MQTT since the early 2000s. He has been involved with the Eclipse Paho open source project since 2011, contributing an MQTT C client. Later, he became Paho's project lead. He worked on the standardization of MQTT 5.0 and is currently co-chair of the OASIS MQTT-SN subcommittee and contributing to the standardization of Eclipse Sparkplug.

Kilton Hopkins started programming computers when he was eight years old. He started a software company a few years later. The world is very different than it was back then, but Kilton is still bringing new technologies to life. Kilton is the co-founder and CEO of Edgeworx, a startup that provides enterprise-scale edge computing. He is the creator of the Eclipse ioFog open source fog and edge computing technology. He is a former IoT advisor to the City of San Francisco and recently served as IoT program director for the Level program at Northeastern University. Kilton lives in Berkeley, California. He received his MBA from the University of Chicago Booth School of Business in 2010.

Kai Hudalla has been working at Bosch on solutions for the Internet of Things in application domains like smart home, energy management, e-mobility, and automotive long before the term reached its recent popularity and ubiquity. He is an active committer on several Eclipse IoT projects, one of which is the Eclipse Hono project, aiming at providing an integration platform for connecting millions of devices to the cloud. Kai also enjoys speaking at conferences, cooking, and golfing whenever weather conditions permit.

Thomas Jäckle is a Software Developer holding a master's degree in application architecture from the Hochschule Furtwangen University in Germany and has been working with Bosch since 2010. Since 2015, he has the position of Lead Software Developer of the Bosch IoT Things cloud service, and since 2017, he is Committer and Project Lead of the Eclipse Ditto project.

Wes Johnson specializes in embedded device software, cloud computing, and the communication technologies that connect them together (IIoT, IoT, M2M, SCADA). He has over 15 years' experience specializing in the industry. After graduating from Portland State University with a BS in computer engineering and physics, he began working at Arcom Control Systems where he was the primary creator of what is now the Eclipse Kura OSGi device framework. At Eurotech, he was Director of Software Engineering and a key contributor to what is now Eclipse Kapua. He is now VP of Software for Cirrus Link Solutions where he co-created the Sparkplug specification and is the project lead for Eclipse Sparkplug and a co-project lead for Eclipse Tahu.

Achim Kraus is a Senior Software Developer and CoAP/DTLS 1.2 CID Evangelist. He is the lead of the Eclipse Californium project. In the past, he worked for Bosch.IO as a Senior Software Developer. Achim obtained a master's in computer science from the University of Stuttgart in 1995.

Ulf Lilleengen is a Software Engineer in the Red Hat Application Services organization, focusing on building the next-generation services for the Internet of Things. Ulf has a passion for open source and has participated in open source communities building software for everything from microcontrollers to servers for the past 15 years. Ulf lives in Hamar, Norway, with his two kids, wife, and two dogs.

Rick O'Connor is President and CEO of the OpenHW Group, a not-for-profit, global organization driven by its members and individual contributors where HW and SW designers collaborate in the development of open source cores, related IP, tools, and SW such as the CORE-V family of open source RISC-V cores. Previously Rick was Executive Director of the RISC-V Foundation, which was launched by him in 2015. Today, under RISC-V International, the RISC-V ecosystem consists of more than 400 members building an open, collaborative community of software and hardware innovators powering processor innovation. With many years of executive-level management experience in semiconductor and systems companies, Rick possesses a unique combination of business and technical skills, and over Rick's career, he was responsible for the development of dozens of products accounting for over $750 million in revenue. Rick holds an Executive MBA degree from the University of Ottawa, Canada, and is an honors graduate of the faculty of Electronics Engineering Technology at Algonquin College, Canada.

Acknowledgments

A tender thank you to Roxanne, my wife. She knew how much of my time this project would require and yet supported me from beginning to end. You truly are the Kaguya to my Miyuki.

Thank you to everyone at the Eclipse Foundation who supported this book. I am grateful for the encouragement provided by Mike Milinkovich, our executive director, and Paul Buck, our VP of ecosystem development and my manager.

Special thanks to Hassan Jaber, my partner in the Eclipse Foundation's marketing team. He was crazy enough to ask his wife to read an early draft of the introduction, and their enthusiastic feedback gave me a much-needed confidence boost.

A round of applause for the technical reviewers who provided feedback on one or several chapters of this book. Some of them even interrupted their vacations to perform their review. Talk about dedication!

Thank you to Jill Balzano and Jonathan Gennick from Apress. Jonathan pitched me the idea of writing this book, and I will forever be grateful for the opportunity.

Finally, how could I forget to mention my dear Ashitaka here? This project is probably the last large one for my trusty Lenovo ThinkStation S30 from 2012.

Introduction

It is 1985. A few weeks ago, my father brought a computer from his office. It is an IBM Portable Personal Computer, model 5155. Portable is relative here; the machine weighed 13.6 kg (30 pounds). My father set it up in the unfinished basement of our home, on an old kitchen table. Even by the standards of the day, the machine is underpowered. The simple BASIC programs I wrote barely kept busy its puny 4.77 MHz 8088 CPU. The built-in 256 KB of RAM was plenty. My IT career was over ten years in the future.

It is 2015. A few weeks ago, I bought my first Arduino and a few sensors. I set up the board on my desk. Compared to my desktop computer, with its 3.2 GHz 6-core CPU and 128 gigabytes of RAM, the little Arduino is hopelessly outclassed. And yet, it is without a doubt the most exciting of the two. The Python programs I write for it are simple. Its 16 MHz 8-bit processor is more capable than the 8088 in the old IBM but can access only 2 KB of RAM. The hardware constraints are severe, but the possibilities are endless.

Between those two moments in my life, the Internet happened. Or rather, it escaped the realms of Defense and Academia to transform our lives completely. At first, its impact was limited to our computers; then, it spread its influence on our phones. Nowadays, it is everywhere. Wherever you go, you will find connected devices that gather data and interact with the physical world. We call this the *Internet of Things* (IoT). This trend is genuine; it is there to stay. This book aims to give you the tools and knowledge your organization needs to capitalize on it.

Why IoT Is Relevant

No business is purely virtual. The Cloud itself is the aggregation of millions of servers spread in data centers located worldwide. Those data centers are under constant monitoring. Sensors for temperature, humidity, air quality, and vibrations ensure the equipment operates in optimal conditions. Surveillance cameras and other security systems keep intruders away. So even if your organization focuses strictly on delivering digital services, it will still benefit from the IoT. The reason for this is simple: our world now revolves around data.

Your organization needs data to make better decisions, whether the decider is artificial intelligence or a human being. Your organization needs data to support the automation of its processes and improve your quality of life at work. Finally, your organization needs data to become more efficient and effective. IoT represents a tool to reach that data. In the words of Marco Carrer, CTO at Eurotech, *"There is a tremendous amount of data that's actually trapped in the field and why IoT matters is because it enables the extraction of data."*[1]

This data, previously, was hard to reach or not reachable at all. What made it accessible is the tremendous affordability and miniaturization of computing devices. My little Arduino, which cost about $20 (Canadian!) in 2015, was an order of magnitude cheaper than the computer I messed around with back in 1985. Moreover, the Arduino is much more power efficient, making it suitable for all sorts of deployment scenarios.

IoT: For All Industries and Use Cases

The proliferation of low-power, affordable, and connected devices means that you can build IoT solutions for a wide variety of industries and horizontal use cases.

From an industry perspective, the 2021 edition of the *Eclipse IoT and Edge Developer Survey*[2] found that Industrial Automation, Agriculture, Building Automation, Energy Management, and Smart Cities were the respondents' top five industry focus areas. These industries have in common the need for automation at a large scale and the potential to benefit from machine learning and artificial intelligence. In a factory, sensors placed on machines can gather the data needed to schedule preventive maintenance and prevent equipment breakdowns, for example. On a farm, video analytics can provide insights into the health of animals, while sensors can detect issues affecting crop growth. The examples are endless.

There are also many IoT horizontal use cases. Edge AI, where developers deploy models on servers as close to the source of the data as possible – if not on the IoT devices themselves – is the most cited example. But control logic, data exchange, and sensor fusion are other important ones. Event stream processing, which can serve in

---

[1] www.youtube.com/watch?v=EiIIG-LUFUc
[2] https://outreach.eclipse.foundation/iot-edge-developer-2021

algorithmic securities trading, fraud detection, location-based services, and many other applications, is another generic use case that significantly benefits from the wealth of data available through IoT devices.

How IoT Projects Are Unique

Most, if not all, IoT projects interact with Cloud-based components and platforms. However, blindly deploying Cloud Native applications using a standard *DevOps* approach outside the confines of the Cloud or the corporate data center is doomed to fail. Why is that? The Edge environment where IoT devices and Edge infrastructure are deployed is widely different from the Cloud at a high level.

The Cloud is homogeneous, centralized, and operates at a large scale; resources are available on demand. On the other hand, the Edge is distributed, heterogeneous, and operates at a small scale; resource availability is limited. In other words, the Edge is the polar opposite of the Cloud, which makes IoT projects wildly different.

IoT projects differ from typical IT projects in other ways:

The timescale of IoT projects spans years and even decades. Replacing all of the sensors in a digital building would be an expensive and lengthy endeavor. Ripping off the machines from a factory floor is even less frequent.

IoT projects involve heterogeneous hardware and software components. There is no "one-stop shop" in IoT. Many integrators serve the market, of course, but doing business with them encapsulates the heterogeneity rather than suppressing it.

IoT components deal with unique constraints. IoT and edge computing nearly always involve ruggedized hardware. When deploying compute, storage, and networking resources outside the data center, they face temperature swings, humidity, dust, and many other dangers. Moreover, many devices need to operate partly or exclusively over battery power. This impacts the software side of things in many subtle ways. For example, vibrations could influence sensor readings, and the IoT device needs to account for that.

Connectivity is a given; stability and reliability are not. The "I" in IoT stands for "Internet." Although not all IoT solutions operate over the public Internet, they all leverage various connectivity technologies. IoT developers need to assume that the network's performance and reliability will vary without warning, which has profound implications on solution design.

What Makes Enterprise IoT Different?

The gap between enterprise IoT and hobbyist IoT is as significant as the gap between IoT projects and other IT projects, if not more.

The wealth of affordable hardware and sensors means that anyone can experiment with IoT at home, especially since many of the software building blocks required are open source. Arduinos and Raspberry Pis are ideal experimentation platforms, given their cost and their rich connectivity. One could even make the case that such devices have replaced yesterday's Commodore 64s and TRS-80 Color Computers as a gateway to an IT career. However, enterprise IoT projects come with specific requirements. Several of those requirements stem from the fact that enterprise IoT often deals with mission-critical and real-time applications.

- **Reliability:** The sensors and microcontrollers used need to face adverse physical conditions over a prolonged period. The devices themselves may be relatively affordable, but the labor costs involved to repair or replace them can be onerous, especially in remote locations. Additionally, some applications may tolerate a level of data loss, but several do not.

- **Sustainability:** Hardware component availability and software component supportability must be compatible with the timescale of the solution.

- **Security:** Security is a growing concern for consumers and organizations alike. The scale of enterprise IoT projects means that security breaches have much more severe repercussions. They can even be life-threatening in applications such as industrial automation and autonomous vehicles.

Open Source: The Best Approach

Enterprise IoT projects must leverage production-grade, commercial-quality hardware and software components, given the requirements involved. Some could see this as an obstacle to the adoption of an open source approach. However, nothing could be further from the truth. At its core, open source is the freedom to access, modify, and redistribute source code.[3] If you need to maintain a solution over years and decades, then the possibility to access the source code is valuable. Wish to integrate components from various vendors? Much more straightforward with the source. Need to tweak a solution's stack for a particular type of application? The source enables you to do so freely. Open source is truly the best approach for IoT.

Above everything else, open source is a business model – and a successful one at that. Over time, any innovation becomes a commodity; the value line is slowly but relentlessly moving up. The consequence is that organizations of all kinds can reduce risk and preserve their capacity to innovate by collaborating and pooling resources to build standard open source components. Such components provide them the proper foundation to bring to market differentiated commercial solutions. In other words, leveraging open source building blocks is a powerful way for organizations to focus their limited resources on competitive endeavors. Open collaboration creates the conditions for successful competition.

About This Book

This book aims to teach you how to build Enterprise IoT solutions using open source components. Specifically, it centers on the many libraries and platforms found at the Eclipse Foundation under the banner of its Eclipse IoT , Edge Native , and Sparkplug working groups. Eclipse IoT, which celebrated its tenth anniversary in 2021, is the industry's largest and most successful IoT open source community. Relevant non-Eclipse open source projects are discussed as well.

Although the book discusses hardware matters, its focus is on software and related open standards. If you are an IoT hobbyist, you will learn to apply your skills in an enterprise context. If you are new to IoT, you will learn about the pros and cons of many established technologies. Both types of readers will understand how to work with the Eclipse open source implementations of the most widespread IoT technologies.

Throughout the book, the concepts will be illustrated through examples of real-world deployments. The examples provided cover a variety of use cases, such as industrial automation, smart agriculture, digital buildings, connected vehicles, and others.

---

[3] It is, of course, much more than that. See the *Open Source Definition* proposed by the *Open Source Initiative* (OSI) for a thorough discussion.

This book contains three main parts:

- **Fundamentals and Protocols:** This part presents the reference architecture used throughout the book as well as the technologies common to constrained devices, edge devices, and IoT platforms.

- **Constrained Devices:** This part discusses picking the hardware and software components you will deploy in the field.

- **Edge Computing and IoT Platforms:** This part explains edge computing itself and why you should rely on it. It also covers edge workload orchestration and gateways, before introducing several IoT Cloud platforms based on microservices that expose a well-defined, stable API. Those platforms are not tied to a specific Cloud provider and are suited to private, hybrid, or public Cloud environments.

Most chapters contain code samples or step-by-step instructions to get you started with the components and platforms discussed. I sincerely hope you will find them helpful.

And now, let's get started!

PART 1

Fundamentals and Protocols

CHAPTER 1

What Is IoT?

Un tas de pierres cesse d'être un tas de pierres dès lors qu'un seul homme le contemple avec, en lui, l'image d'une cathédrale.

A rock pile ceases to be a rock pile the moment a single man contemplates it, bearing within him the image of a cathedral.

—Antoine de Saint-Exupéry, *Pilote de Guerre (Flight to Arras)*

IoT is the culmination of a centuries-long evolution. The origins of innovations like the wheel and simple machines are today forgotten, lost in the mists of prehistory. Humans have harnessed the power of water and wind early. Historians trace the origins of the waterwheel back to ancient Greece, over 3000 years ago; as for the windmill, they started appearing as we know them today around the 8th and 9th centuries in the Middle East and Western Asia. In 1712, Thomas Newcomen developed the first commercially successful steam engine. The 18th century was thus the century of mechanization. The 19th and 20th centuries saw two fundamental technology transitions: electrification and computerization. By the 1980s, computers started to take over both factories and offices.

Although we can trace its roots back to the 1960s, the Internet took flight in the early 1990s. It gradually transformed from a document-based platform into a dynamic application platform. However, the advent of affordable wireless data connectivity at the same time – mobile broadband Internet access appeared with the second generation (2G) of mobile connectivity in 1991 – was the decisive advance that made the IoT possible. 2G, along with Wi-Fi networks and Bluetooth, was a game changer. And while the term "Internet of Things" was coined by Kevin Ashton in 1999,[1] many are still unsure about its meaning. My goal is to change that.

---

[1] `www.smithsonianmag.com/innovation/kevin-ashton-describes-the-internet-of-things-180953749/`

© Frédéric Desbiens 2023
F. Desbiens, *Building Enterprise IoT Solutions with Eclipse IoT Technologies*,
https://doi.org/10.1007/978-1-4842-8882-5_1

To achieve this goal, I will first explain what IoT is and how it is different from other current technology trends. Second, I will define edge computing and how IoT can leverage it. I will then introduce an IoT reference architecture that spans constrained devices, edge nodes, and IoT platforms. Finally, I will discuss IoT protocols, which play a horizontal role in the architecture and bring the other components together.

IoT and Other Current Trends

Before going any further, let's try to define IoT itself. Many different definitions are floating around. Here are a few which I find meaningful:

- **Gartner:**[2] *The Internet of Things (IoT) is the network of physical objects that contain embedded technology to communicate and sense or interact with their internal states or the external environment.*

- **Wikipedia:**[3] *The Internet of Things (IoT) describes physical objects (or groups of such objects) that are embedded with sensors, processing ability, software, and other technologies, and that connect and exchange data with other devices and systems over the Internet or other communications networks.*

- **Alexander S. Gillis:**[4] *The Internet of Things, or IoT, is a system of interrelated computing devices, mechanical and digital machines, objects, animals, or people that are provided with unique identifiers (UIDs) and the ability to transfer data over a network without requiring human-to-human or human-to-computer interaction.*

While they are a good starting point, the definitions reproduced before do not catch some nuances I feel are essential. Here are the three main nuances missing from those definitions:

- **The I in IoT means "Network":** Connectivity is a given when working with the IoT. However, not all deployments rely on carrier networks or the public Internet. Many narrowband technologies can integrate directly to a local wired or wireless network without Internet connectivity.

---

[2] www.gartner.com/en/information-technology/glossary/internet-of-things
[3] https://en.wikipedia.org/wiki/Internet_of_things
[4] https://internetofthingsagenda.techtarget.com/definition/Internet-of-Things-IoT

- **This is not just about sensors:** Most IoT devices can count on an array of sensors to report the state of the physical world. They can do much more than just gathering data, however. When equipped with actuators, IoT devices can accept commands and interact with their environment. They can turn on lights, adjust the position of a valve, and increase the speed of a ventilator; the possibilities are endless.

- **There is a human element:** IoT devices exist in the physical world, as humans do. There are several use cases where adding human user interface controls to a device makes sense. In some instances, such as autonomous forklifts and trucks used in warehouses or around industrial sites, workers can even use the controls to take over the vehicle. Orderly coexistence between humans and devices will become increasingly important.

Given the points I make previously, I feel the following definition is probably the best compromise:

The Internet of Things (IoT) is a system of networked physical objects that contain embedded hardware and software to sense or interact with the physical world, including human beings.

Now that we have a clear understanding of the IoT, let's disambiguate it from a few related concepts.

- **M2M (machine to machine) communication:** This term has been in use for a while; in fact, Eclipse M2M was the initial name of the Eclipse IoT working group. The rise of industrial computing saw machines being connected through point-to-point connections initially. Nowadays, industrial networks are commonplace in factories. M2M is a concept rooted in industrial automation and closely related to operational technology.

- **Web of Things:** The Web of Things (WoT) is a set of standards from the World Wide Web Consortium (W3C) that aim to enable "easy integration across IoT platforms and application domains." WoT is strictly focused on software and much narrower in scope than IoT. The Eclipse Thingweb project is the official reference implementation of the W3C WoT Working Group.

- **Industry 4.0:** The ongoing digitalization of the industrial sector has led some to consider that a fourth industrial revolution is underway. The "Industry 4.0" moniker refers to that. M2M and IoT represent technological approaches that make possible the automation of traditional manufacturing and industrial processes. Ultimately, Industry 4.0 will lead to the widespread adoption of intelligent machines capable of making their own decisions through artificial intelligence (AI).

- **Industrial Internet of Things (IIoT):** This one is relatively straightforward. IIoT is the application of IoT technologies to industrial applications, such as manufacturing, asset tracking, and energy management. IIoT solutions usually replace legacy control systems intending to improve productivity and efficiency.

Ultimately, IoT devices must send their data somewhere; the commands they receive must have an origin. The Cloud is typically both the destination of the data and the source of the commands; whether this Cloud is private, hybrid, or public is irrelevant. However, connecting IoT devices directly to the Cloud is not always possible or desirable. For this reason, contemporary IoT solutions often rely on edge computing.

Edge Computing and IoT

Edge computing is a form of distributed computing that brings compute, storage, and networking resources closer to the physical location where the data is produced and commands are executed. Its origins go back to the early days of computing. For example, the consoles used by the mission controllers during the Apollo era were not displaying live data streamed from the spacecraft. Three antennas in Australia, Spain, and the United States captured the telemetry and crunched it locally on Univac 1230 mainframes before forwarding it to the IBM S/360 driving the consoles in Houston. Figure 1-1 illustrates the architecture.

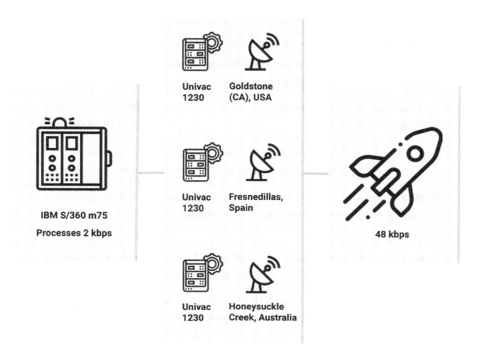

Figure 1-1. *IT architecture of the Apollo Moon Missions (Credit: Eclipse Foundation)*

What makes edge computing different nowadays is that it does not rely directly on the Cloud Native and DevOps approaches but rather on adaptations to the Edge environment such as Edge Native and EdgeOps.

For the time being, I would like to explain why edge computing technologies are so crucial to successful IoT implementations. Out of the many possible reasons, I think the following ones are the most compelling:

- **Optimize bandwidth usage:** The emergence of 5G and affordable narrowband connectivity options – such as Dash7, LoRaWAN, and Zigbee – makes the deployment of IoT devices where they are needed simpler and more affordable. Many IoT use cases generate gigabytes or even terabytes of data every day. Transmitting the integrality of that data does not make sense from a cost perspective.

- **Reduce latency:** The latency on most corporate and industrial networks is highly variable. Reaching to the Cloud over the public Internet adds a level of unpredictable latency. Deterministic local

7

area networks provide a way to make latency predictable in a specific location, but Internet service providers cannot deliver that level of predictability. Low latency is vital to mission-critical and real-time uses cases. Suppose you are automating a power plant, a factory, or a self-driving vehicle. In that case, you cannot afford to have time-sensitive decisions made in the far-away Cloud due to the latency induced by the Internet.

- **Support data sovereignty:** Data is subject to several layers of laws and regulations where it is collected, stored, and even processed. Data sovereignty describes the efforts made by several national or regional governments to protect the data of their citizens and corporations. In highly regulated industries, such as healthcare, this translates into encryption and data residency requirements; the data must be encrypted on the move and at rest and stored inside a specific geographical area. This area can be a building, a city, or maybe a state or province in the case of countries with a federal structure. Edge computing provides infrastructure that makes it easier for solutions to implement data sovereignty and data residency requirements.

- **Implement reporting by exception:** The value of the data gathered by devices in IoT solutions is not always the same. It is essential to process every reading or archive it in a historian database for future reference for some use cases. For other use cases, the only readings that matter are those outside the normal thresholds of operation. In other words, only exceptions are reported. Developers can implement reporting by exception directly on the device driving the sensors. However, in scenarios involving complex data models, machine learning, and video analytics, it is better to offload such decisions to an edge node in the vicinity. Reporting by exception is especially useful when devices operate on battery power since it reduces the need to send data over the network, which is always a power-hungry operation.

The takeaway of this section is this: edge computing is highly beneficial to most IoT solutions but the simplest one. This explains why the reference architecture places edge computing front and center.

IoT Reference Architecture

A large part of my job at the Eclipse Foundation is introducing newcomers to our IoT and edge computing technologies. At the time of writing, this represents over 50 open source projects of all shapes and sizes. To help developers understand what is available, our community put together an IoT reference architecture, reproduced in Figure 1-2.

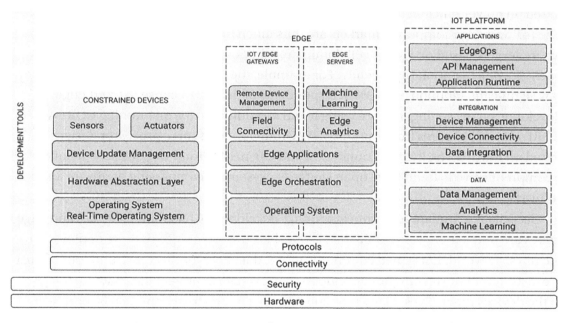

Figure 1-2. *IoT reference architecture (Credit: Eclipse Foundation)*

One important fact about this diagram is that it does not represent a blueprint. The Eclipse community is a code-first community where developers come together to build components and platforms. The reference architecture is a map to the ecosystem they built, not some grandiose architectural vision that our committers and contributors followed. That said, let's have a closer look at each of the significant parts of the architecture and the related Eclipse open source projects.

Common Layers

The reference architecture defines three types of environments: constrained devices, edge, and IoT platforms. Most of the concepts exist in the scope of these three environments. However, the architecture contains five concepts that apply across the three environments. Those are hardware, security, connectivity, protocols, and development tools.

Hardware strongly influences the design of IoT solutions, especially when considering constrained devices and edge nodes. IoT developers optimize their code for size most of the time and strive to keep power consumption to a minimum – at least for devices operating on battery power. Device designers usually pick microcontrollers based on power-efficient processor architectures such as Arm for the same reason. The IoT and edge computing markets are thus much more diverse than the Cloud one, where servers based on x86-64 still dominate. Additionally, open source hardware has been growing in popularity recently. For example, the open source RISC-V architecture saw significant adoption, with 9% of respondents of the 2021 Eclipse IoT and Edge developer survey stating they are using it and an additional 8% mentioning CORE-V. The CORE-V family of cores from the OpenHW Group is open source processor designs based on RISC-V. The Eclipse Foundation and OpenHW Group work closely to build a comprehensive IoT and Edge open source ecosystem encompassing hardware and software.

Security is a fundamental concern for IoT developers. Constrained devices and edge nodes are much more vulnerable than Cloud-based servers since they are deployed in the physical world. Encrypted communications, data encryption at rest, and root of trust are just a few of the techniques developers can leverage to protect data and the integrity of the devices themselves. Many now consider that a zero-trust approach is more secure since it implies that software never trusts devices by default. There are many aspects to security, and no single platform or open source project can cover them all. Sticking to active, well-maintained projects that keep their dependencies up to date will help you mitigate some of the risks.

Connectivity refers to the wired or wireless networking technologies used by a solution. The 2021 Eclipse IoT and Edge developer survey found that most developers leverage mature connectivity options such as Ethernet, Wi-Fi, and Bluetooth. The growing availability of 5G coverage will probably change things soon. Moreover, many narrowband technologies such as Dash7, LoRa, and Sigfox are well suited to IoT projects involving sensor data collection. Zigbee and Z-Wave are also worth mentioning since they use only a fraction of the power required by Wi-Fi. Support for most specialized connectivity options is usually implemented in a gateway that connects to the Internet – and thus the Cloud – through a more traditional connection.

Protocols play a critical role in most IoT architectures. Since there is much to say about them, I will cover them in greater detail later. Always keep the dessert for last!

Finally, development tools are an essential concern for the Eclipse IoT and Edge community. The traditional Eclipse Integrated Development Environment (IDE) is still going strong with over six million users. It is widely popular for embedded development, which makes it a strong choice for IoT developers. The Eclipse CDT and Eclipse Embedded CDT projects are proof of that, as is the CORE-V IDE built by the OpenHW Group team. However, the global trend toward Cloud-based development environments is also growing in the IoT and Edge market. Eclipse Che, a Kubernetes-Native browser-based IDE for developer teams, has seen significant adoption with IoT developers in the last few years. Eclipse Theia, an extensible cloud and desktop IDE Platform based on web technologies, is leveraged by Che and a growing set of products such as the Arduino IDE. Developers can also use a generic version of Theia through Eclipse Theia Blueprint, a template for building desktop-based products based on the Eclipse Theia platform. It is worth mentioning that very few of the technologies hosted by the Eclipse IoT and Edge Native working groups have ties to a specific IDE.

Note   Do not confuse the Eclipse IDE and the Eclipse Foundation. IBM open-sourced the Eclipse IDE in 2001. The participants to the project created the Eclipse Foundation in 2004 to be the vendor-neutral steward of the Eclipse IDE. Over time, the Eclipse Foundation diversified and created several working groups covering multiple technology areas. The Eclipse IoT working group, for example, celebrated its 10th anniversary in 2021.

As of writing, the Eclipse Foundation hosted over 425 open source projects and had four strategic areas: automotive, Cloud Native Java, IoT and edge computing, and development tools.

Constrained Devices

Constrained devices take us back, in a way, to the earlier stages of computing. 32-bit processors are still typical in them, and they typically have little memory and storage. What makes them unique is their capacity to operate on battery power for months or even years and their rich input and output capabilities. Embedded developers have programmed constrained devices for a very long time. What makes IoT constrained devices different is that they can connect to the Internet or isolated networks using Internet-class technologies.

While some developers still deploy their constrained device code on the bare hardware, most of them leverage an *operating system* (OS) or *real-time operating system* (RTOS). There are several options available in the market, both commercial and open source. The most popular open source options in the 2021 Eclipse IoT and Edge developer survey were FreeRTOS, Mbed OS, and Zephyr. Linux is also widely adopted, although it requires more powerful processors and more memory. For developers, leveraging an OS or RTOS makes much sense since they implement generic low-level features that their applications can leverage.

Note   RTOSes are different from traditional operating systems in two ways: they are predictable and deterministic. Predictable means that the system will respond to a specific event within a strict time limit. Deterministic means that the behavior of the RTOS scheduler can be predicted. In other words, we know how much time an OS-level operation will take (predictable) and that the result will always be the same (deterministic).

While operating systems provide developers with a standard application programming interface (API), working with sensors and actuators often requires custom code. Moreover, porting an application from one microcontroller to another – or even just migrating to a different revision – can be cumbersome. This is where hardware abstraction layers (HAL) come into play. HAL provide a hardware-independent and OS-independent API that makes applications easier to port. Since the life cycle of IoT solutions is long, the probability that you will need to port your code to different hardware or a new version of the RTOS you use is high. Portability is thus a significant concern. In the Eclipse IoT toolkit, the Eclipse MRAA project provides abstractions for various boards' I/O pins and buses (such as the Raspberry Pi), while Eclipse UPM interfaces with sensors and actuators.

> **Note**   Some operating systems are pushing abstraction layers to the next level by enabling developers to pick the kernel that best fits their use case and by providing uniform higher-level APIs to applications. An excellent example of this approach is the Oniro project. Oniro can use the Linux, Zephyr, or LiteOS kernels. It possesses a kernel abstraction layer that uniformizes process and thread management, memory management, file system, network management, and peripheral management. Oniro is a compatible implementation for the global market of OpenHarmony, an open source operating system specified and hosted by the OpenAtom Foundation.

A critical issue with IoT constrained devices is to keep them up to date. Over time, vulnerabilities will appear in the code you wrote and the OS you use. Malicious actors are ready to take advantage of those vulnerabilities to conduct cyberattacks.[5] If you are an IoT developer or aim to become one, it is essential to take security updates seriously. Some RTOSes and protocols feature built-in update management that will integrate with a variety of server-side platforms. Eclipse hawkBit is an example of such a platform. The Eclipse Hara project provides a sample client implementation of the hawkBit APIs that you could leverage – or at least take inspiration from – on your constrained device. Some will say, and rightfully so, that a zero-trust approach will help mitigate the dangers posed by compromised devices. However, malicious actors could use such devices to attack others, not just your organization. Consequently, having a responsible security posture is vital.[6] This includes offering at least one channel for your customers and end users to report security vulnerabilities and committing to act once the reports are in your hands.

Lastly are sensors and actuators. Picking the right components can be a complicated task. In the case of sensors, their precision, durability, and sampling rates are all critical considerations. As for actuators, they accept an electric signal and combine it with a source of energy. There are several varieties, including electric, hydraulic, and pneumatic, among others. Most of the boards featuring microcontrollers available in the market offer an array of built-in sensors, plus I/O and buses to connect sensors and actuators.

---

[5] https://www.zdnet.com/article/your-insecure-internet-of-things-devices-are-putting-everyone-at-risk-of-attack/

[6] https://www.zdnet.com/article/the-iot-is-getting-a-lot-bigger-but-security-is-still-getting-left-behind/

Edge

I described edge computing earlier as a form of distributed computing. However, not all distributed computing is edge computing – even in an industrial context. Modern edge computing emphasizes the use of microservices deployed in containers or, less frequently, virtual machines. It also relies on approaches and techniques inspired by *DevOps*. Simply having software running on a computer connected to a *programmable logic controller* (PLC) driving the machinery is not edge computing.

There are two main types of edge computing nodes: gateways and servers. Edge servers can run multiple workloads and represent a way to perform workload consolidation. They can be used to connect to constrained devices or other edge servers. Gateways, on the other hand, are further classified into IoT gateways and pure edge gateways. An IoT gateway acts as the focal point for a group of sensors and actuators. It provides connectivity to these devices to each other and an external network. It can be a physical piece of hardware or embedded in a network-connected device. An edge gateway is the connection point between an edge server and the Cloud or a more extensive network.

Independently of their type, edge nodes usually run an operating system. The 2021 Eclipse IoT and Edge developer survey found that Linux largely dominates on edge nodes, with Microsoft Windows a strong second. Initiatives such as EVE-OS, hosted at LF Edge (Linux Foundation), aim to bring consistent system and orchestration services to this space on the top of the Linux kernel. The Oniro project, which I previously mentioned, is also suitable for edge nodes.

Since edge computing relies on microservices, it is necessary to configure, manage, and coordinate them automatically. This is *edge orchestration.* While automation is usually about a single task, orchestration focuses on processes that involve multiple steps across several services. In the Cloud space, this need for orchestrating workloads has increasingly been fulfilled by *Kubernetes* (also known as K8s). However, the complexity and resource requirements of Kubernetes do not make it a good fit for several edge use cases. Some open source projects address this issue. K3s is a lightweight distribution of Kubernetes, packaged as a single binary, that minimizes external dependencies. KubeEdge, on the other hand, provides a containerized edge computing platform completed by a Cloud component that integrates with Kubernetes' API server. That said, a growing part of the edge developer community prefers to leverage edge platforms that can integrate with Kubernetes when needed but can operate in stand-alone mode. This is the case of the two edge orchestration projects calling the Eclipse

Edge Native working group home. Eclipse ioFog focuses on container orchestration at the edge and represents a centralized take on edge computing since edge nodes communicate with a central controller. Eclipse fog05 (pronounced fogOS; the 05 is for 5G) supports containers, virtual machines, and even binaries (executables); it manages nodes in a decentralized fashion by incorporating resources in a unified fabric.

Note   For a thorough discussion of whether Kubernetes is a good fit for your project, you can watch a talk titled "Do We Really Need Kubernetes at the Edge?" which I delivered in 2021 (`https://youtu.be/u8BDtSP7Dfg`).

Edge orchestration platforms aim to enable the deployment of edge applications. Such applications come in all shapes and sizes, cover a wide variety of use cases, and apply to nearly all industries. The Eclipse IoT working group hosts a few open source edge applications. Eclipse 4diac provides development tools and a runtime environment compatible with IEC 61499; in other words, it is a complete toolkit to build bespoke programable logic controllers to drive industrial equipment. Eclipse Keyple is a widely adopted solution to build complex contactless ticketing, transportation, and event access systems. If you ever took the subway or the bus in Paris or bought an SNCF train ticket from a terminal, you used Keyple without even knowing it. Finally, Eclipse VOLTTRON is a distributed control and sensing software. VOLTTRON analyzes and converts digital building data streams into actionable information to improve building operations, manage energy consumption, and enable building integration with the electrical grid.

IoT/edge gateways and edge servers have distinct feature sets given their respective role in the architecture. Gateways are a bridge between constrained devices and other resources found on the edge-to-cloud continuum. In addition to modern field connectivity options, they often feature serial ports and other types of legacy interfaces, enabling organizations to leverage existing assets in the context of the IoT. Because gateways are a connection point for constrained devices, they frequently offer remote device management features. One of the longest-running Eclipse IoT projects, named Eclipse Kura, is a platform for building Linux-based gateways with the Java language and the OSGi framework. Kura offers API access to the hardware and supports multiple field protocols, including Modbus, MQTT, OPC UA, and S7. Kura can run edge applications and orchestrate the workloads, making it a comprehensive solution for gateways.

While gateways play a specific role in the architecture, edge servers provide the resources required to execute workloads. The diagram identifies machine learning and edge analytics as the two most critical. The whole story is slightly more nuanced, however. The respondents of the 2021 Eclipse IoT and Edge developer survey identified artificial intelligence, control logic, data analytics, and sensor fusion as their top edge computing workloads – in that order. One way or another, edge servers are a powerful tool to reduce latency for artificial intelligence and machine learning while improving privacy, especially for solutions deployed in the public space.

IoT Platforms

If constrained devices gather data and edge nodes process it in physical locations close to its source, one could say that IoT platforms manage and analyze data. However, this statement expresses only a part of the scope of IoT platforms. They also provide the infrastructure to run applications that leverage the data and integration features that link constrained devices and edge nodes to the rest of the infrastructure managed by corporate IT.

The first group of concepts related to IoT platforms in our architecture pertains to data. Data is integral to IoT. Sensors create it, and actuators execute commands that are the consequence of data-driven decisions. In that context, machine learning (ML) is an essential capability for IoT platforms since it aims to transform raw data into knowledge. Developers can leverage many popular open source machine learning frameworks, such as TensorFlow and scikit-learn. Some ML frameworks have even been designed from the ground up for deployment at the edge; this is the case with TensorFlow Lite and Apache MXNet. There are several ML-related projects at the Eclipse Foundation. The most mature and widely adopted is Eclipse Deeplearning4j. Deeplearning4j is a suite of tools for running deep learning on the JVM. It allows developers to train their models using the Java language and provides interoperability with the Python ecosystem.

Analytics is another word to describe the use of data analysis tools and procedures. There is a thriving market of solutions in that space, both proprietary and open source. Bruce Sinclair distinguishes three classes of analytics:[7]

- **Real time:** Analytics performed as soon as the data is acquired to bring anomalies to light.

---

[7] See this episode of the "The IoT Inc Business Show" podcast for more details. https://www.iot-inc.com/three-main-classes-internet-of-things-data-analytics-podcast/

- **Predictive:** Analytics looking at current and historical data to produce a prediction and confidence level.

- **Descriptive:** Analytics that report on past, present, or future data and are often the basis of visualizations.

Eclipse Streamsheets is a platform to access, create, and manage server-based spreadsheets that consume, process, and produce real-time data streams. Those spreadsheets can contain several types of graphs in addition to numerical data, enabling users to create interactive dashboards. Streamsheets supports several IoT protocols out of the box.

Data management considers data a valuable resource and provides governance, tools, and methods to access, integrate, cleanse, and store it. Data management is a vast functional domain covering database management systems, big data systems, data warehouses, and data lakes. Data modeling is also a paramount concern since models are helpful abstractions developers leverage when building software. The Eclipse Kapua project provides a modular IoT platform with solid data management features. Kapua users can also perform IoT analytics by building dynamic dashboards.

Let's now focus on integration, the second group of concepts specific to IoT platforms in our reference architecture. The three concepts in this group are data integration, device connectivity, and device management. It is worth mentioning that Eclipse Kapua implements all three concepts and offers streamlined integration with gateways running Eclipse Kura.

Data integration is the process of creating unified sources of information from multiple data sources. The wide variety of data management technologies and approaches makes data integration a critical function in the architecture. Of course, models and data sharing frameworks are also an essential part of data integration since it is rather challenging to consolidate data if it is not expressed in a standardized and consistent way. The Eclipse Dataspace Connector project provides an interoperable framework for data sharing built on a common identity model. It will implement the International Data Spaces standard (IDS) and the GAIA-X European project's protocols. Another relevant project is Eclipse Thingweb, an implementation of the *Web of Things* (WoT) model proposed by the World Wide Web Consortium (W3C). WoT aims to provide standard metadata and APIs to make possible integration across IoT platforms and application domains. If WoT is your thing, you may also want to look at Eclipse EdiTDor, a tool to create and edit WoT artifacts that is available online.

While networks and protocols provide a basic level of device connectivity, deploying an IoT solution at scale typically requires connecting to many constrained devices. Moreover, those devices will likely come from different providers and will not necessarily use the same protocols – whether by constraint or because of a design decision. Eclipse Hono provides remote service interfaces to interact with IoT devices in a protocol-agnostic way. Devices can send *telemetry* and *event* messages to Hono to report sensor readings; applications can send *command* messages to the devices. Behind the scenes, Hono supports the AMQP, CoAP, HTTP, and MQTT protocols, and developers can build custom protocol adapters as needed.

At its core, device management covers the operation and maintenance of devices. An IoT platform needs to maintain a device registry and record each device's update history to achieve this. Since IoT devices are so varied in their hardware, configurations, and capabilities, many of the features of device management platforms used to control computers and mobile devices are not needed in the context of IoT. However, a robust software update delivery back end is a must. The most advanced solutions support partial downloads and enable devices to resume interrupted file transfers. Eclipse hawkBit, a domain-independent back-end framework for rolling out software updates to constrained devices, provides that and much more. In particular, hawkBit's Device Management Federation (DMF) API makes it possible to leverage the device management features in protocols that include them, such as Lightweight M2M (LwM2M).

Ultimately, applications are the reason why IoT platforms need to gather data and provide integration capabilities. Deploying constrained devices in the field is a means to an end, providing the infrastructure required by applications. The Eclipse ecosystem provides application runtimes and API management capabilities. Moreover, the ioFog and fog05 edge computing platforms implement the core principles of EdgeOps, an evolution of the DevOps approach specifically targeted at edge environments.

The 2021 Eclipse IoT and Edge developer survey found that for IoT platforms, the top three languages favored by developers are Python (22%), Java (18%), and JavaScript (16%). But what about runtimes? In the case of Python, the Python interpreter is the runtime; for JavaScript, Node.js is the most popular one. But what about Java? In the last few years, the Java ecosystem saw the Eclipse Foundation become its nexus. This started with the creation of the Microprofile project, which is a platform definition that optimizes Enterprise Java for a microservices architecture. The trend continued with Oracle contributing Java Enterprise Edition and other technologies to Eclipse; they are now thriving under a new name: Jakarta EE. At the end of 2020, the OSGi Alliance announced it would transition to the Eclipse Foundation. OSGi is a popular dynamic module system for Java that has seen significant adoption in the IoT market, with Eclipse Kura as one of the most prominent adopters. Then, in early 2021, the AdoptOpenJDK initiative became the Adoptium working group at Eclipse. Eclipse Temurin is the name of the OpenJDK distribution from Adoptium. The Eclipse Foundation is the home of Cloud Native Java, and several of the IoT platforms I already discussed use the technology.

Note   A common misconception about Eclipse open source projects is that the Foundation mandates using the Java programming language and the Eclipse Public License (EPL). There are no such restrictions. Projects are free to leverage the languages of their choice. As for licensing, while using the EPL v2.0 is possible, projects can also pick the Apache 2.0, MIT, and BSD licenses if they wish. Dual licensing is also possible, and the EPL v2.0 even supports using the General Public License (GPL) version 2.0 or later as a secondary license.

From a general perspective, API management is a way to create, publish, and monitor the runtime performance of service interfaces. In most cases, API management platforms enforce usage policies and access control. In the last few years, the concept of digital twins has grown in popularity in the IoT market. A digital twin is simply a virtual representation of a physical device or asset updated from real-time data. The Eclipse Ditto project provides synchronous and asynchronous APIs to interact with digital twins and performs resource-based access checks on each API call. Ditto is thus an API management platform focused on digital twins. Moreover, it performs state management; Ditto keeps track of the reported, desired, and current state of the devices it monitors and takes care of publishing state changes.

Finally, there is EdgeOps. Maybe you are asking yourself why we needed such a thing when DevOps transformed how we build, deploy, and monitor software. The answer is simple. A pure DevOps approach will not work outside the Cloud or the corporate data center. Of course, containerized microservices are essential at the edge, as is infrastructure as code. The close collaboration between development and operations is also a huge advantage, whether your organization merged the teams or not. However, many IoT and Edge deployments have strict real-time requirements or assume a mission-critical role. In that context, continuous integration and delivery (CI/CD) must consider the environment's constraints. This fact explains why I wrote that EdgeOps is an adaptation of DevOps to the edge earlier. What makes ioFog and fog05 so powerful as edge computing platforms is that their contributors built them from the ground up for the edge. They are an embodiment of EdgeOps.

There is one Eclipse project I didn't mention yet that you should be aware of: Eclipse Arrowhead. Arrowhead is a framework enabling developers to build, design, implement, and deploy automation Systems of Systems. From an IoT point of view, you can use it to orchestrate the various components of an IoT platform. It even provides several valuable services, such as an event handler, a device registry, and a device manager.

Now that we have completed our tour of the reference architecture, let's circle back to protocols. They are what brings constrained devices, edge nodes, and IoT platforms together.

Protocols: Foundational Building Blocks

Developers coming to IoT and Edge from the Cloud world sometimes assume that HTTP and REST, those Swiss knives of the modern microservice builder, are all they need. This is, of course, a mistake. IoT and Edge devices operate in an environment where reducing battery drain and bandwidth usage is vital. Because of that, many solutions use UDP as their transport rather than TCP (although many IoT-specific protocols support both). There is also a greater variety of topologies and interaction models at the edge. Device meshes are frequent in IoT projects, and the same is true of publish/subscribe interactions. The consequence of this is that there are many IoT-specific protocols available. Some of them, such as MQTT and OPC UA, have even been around for over 20 years. Compared to the protocols traditionally used in industrial automation, such as Modbus and Profinet, modern IoT protocols are more secure and can be routed safely over the public Internet. Consequently, most Cloud IoT middleware supports them.

Current IoT protocols leverage one of two main interaction models: request/response or publish/subscribe. In the request/response model, a client establishes a connection to a specific server, sends a request, and eventually gets a response – synchronous or not. HTTP is a typical example of the request/response model. In the publish/subscribe model, senders (publishers) are entirely decoupled from receivers (subscribers). Publishers send messages to a specific location, and subscribers will receive messages sent to that location if they register to get them. As a result, publishers are not aware if there are subscribers or not. MQTT is the most widely used implementation of the publish/subscribe model.

At the time of writing, there were 11 IoT and edge protocol implementation projects at the Eclipse Foundation. Eight of them implement standards under the jurisdiction of external standards bodies. They are listed in Table 1-1.

Table 1-1. *Implementations of protocols under external jurisdiction*

Protocol	Organization	Model	Eclipse Projects
Constrained Application Protocol (CoAP)	Internet Engineering Task Force (IETF)	Request/Response	Eclipse Californium
Data Distribution Service (DDS)	DDS Foundation (Object Management Group)	Publish/Subscribe	Eclipse Cyclone DDS
Lightweight M2M (LwM2M)	OMA SpecWorks	Request/Response	Eclipse Leshan Eclipse Wakaama
OPC UA	OPC Foundation	Request/Response Publish/Subscribe	Eclipse Milo
MQTT	OASIS Open	Publish/Subscribe	Eclipse Amlen Eclipse Mosquitto Eclipse Paho

In addition, there are three protocol implementations completely under the governance of the Eclipse Foundation. Those are detailed in Table 1-2.

Table 1-2. *Protocols under the governance of the Eclipse Foundation*

Protocol	Model	Eclipse Projects
Production Performance Management Protocol (PPMP)	Data format only	Eclipse Unide
Sparkplug	Publish/Subscribe	Eclipse Sparkplug Specification Eclipse Tahu
zenoh	Publish/Subscribe	Eclipse zenoh

Of those three, only Sparkplug is managed under the Eclipse Foundation Specification Process for the time being. This process, which complements the Eclipse Development Process, defines the open source rules of engagement, the organizational framework, and the workflow for specification projects at the Eclipse Foundation. Eclipse specification projects are working in the open in a transparent, vendor-neutral fashion. Anyone can propose changes to the specifications by opening an issue or submitting a pull request on the specifications' public repository. Whether those proposals will be considered or not is up to the committers of the specification project.

With so many options to choose from, picking the most suitable protocol for your solution could look like a daunting task. Here are a few questions to ask yourself to guide your choice:

- **What am I trying to achieve?** If you are simply gathering sensor data, a publish/subscribe protocol is probably the best option. Request/response protocols are a better fit for commands – at least if you have real-time requirements.

- **What are my constraints?** Some protocols are a better fit to optimize throughput; others help in reducing latency or power consumption. Do not blindly trust the assertions of others and test yourself!

- **What is my ecosystem?** You could pick sensors, actuators, or other hardware and then use the protocols they support. Conversely, you could pick a protocol and then shop for suitable products. Either way, you will define the boundaries of your ecosystem.

An Extensive IoT and Edge Toolkit

Our tour of the reference architecture and the Eclipse Foundation's IoT and Edge ecosystem ends here. As you can see, the toolkit at your disposal is quite extensive. Arnold Vogt, Lead IoT and Digital Transformation Analyst at Pierre Aucoin Consultants, wrote about it in a report: "From our point of view, the Eclipse Foundation's IoT Working Group is undoubtedly the leading open source IoT community. It's no surprise to us that our two best-in-class vendors are part of that community."[8]

---

[8] You can download the report, titled *Open Digital Platforms for the Industrial World in Europe 2021, here:* https://outreach.eclipse.foundation/pac-radar-research-iot-open-source

But you don't have to take my word for it or even Arnold's word for that matter. Figure 1-3 shows our reference architecture with the logos of the projects discussed earlier.

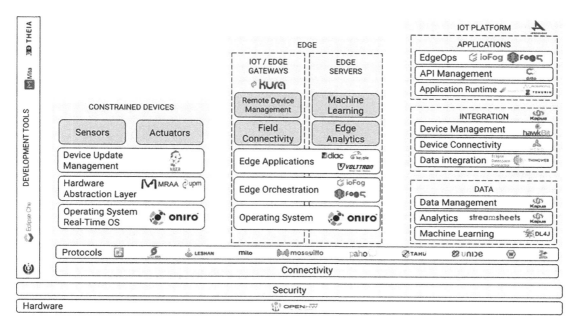

Figure 1-3. *IoT reference architecture with project logos (Credit: Eclipse Foundation)*

Seeing is believing. Our IoT and Edge community built not just a rock pile but a cathedral of code.

CHAPTER 2

CoAP

Une civilisation se construit par l'apport successif de générations prenant appui l'une sur l'autre comme les pierres d'un édifice.

A civilization is built by the successive contribution of generations leaning on each other like the stones of a building.

—André Frossard, *Le Monde de Jean-Paul II*

Open source is about leveraging building blocks implementing shared functionalities. This enables you to focus on your use case instead of reimplementing what countless others have already implemented. Since your time is a severely constrained resource, any minute you save by using open source components is a minute you can invest in meaningful differentiation for your solutions.

Open source projects and standards also benefit from the approach I just described. When an open source project adopts another as a dependency, it reinforces their respective communities. Having an open source adopter is a valuable source of feedback and new requirements for the dependency. For the adopter, leveraging an open source dependency provides flexibility and control proprietary components are unable to provide. Several of the Eclipse IoT and Edge building blocks depend on their peers.

This chapter and the next cover two closely related protocols: CoAP and LwM2M. In this one, I will also teach you how to use Eclipse Californium, a mature CoAP implementation; in the next one, you will learn about Eclipse Leshan, a framework to build LwM2M solutions using Californium as its CoAP implementation.

© Frédéric Desbiens 2023
F. Desbiens, *Building Enterprise IoT Solutions with Eclipse IoT Technologies*,
https://doi.org/10.1007/978-1-4842-8882-5_2

CoAP: A Slimmer HTTP

Like HTTP, the *Constrained Application Protocol* (CoAP) is described in a *Request for Comments* (RFC) published by the *Internet Engineering Task Force* (IETF). At the time of writing, CoAP is a proposed Internet standard. The initial RFC for CoAP, RFC 7252, published in June 2014, has been updated by the following RFCs:

- **RFC 7641:** *Observing Resources in the Constrained Application Protocol (CoAP)*, September 2015

- **RFC 7959:** *Block-Wise Transfers in the Constrained Application Protocol (CoAP)*, August 2016

- **RFC 8613:** *Object Security for Constrained RESTful Environments (OSCORE)*, July 2019

- **RFC 8974:** *Extended Tokens and Stateless Clients in the Constrained Application Protocol (CoAP)*, January 2021

All four represent optional functionality that may not be present in specific implementations.

It is worth mentioning that encrypted CoAP connections rely on the *Datagram Transport Layer Security* (DTLS) protocol version 1.2, published originally in January 2012 as RFC 6347. DTLS is based on the *Transport Layer Security* (TLS) protocol commonly used to secure HTTP connections but focuses on datagram protocols such as the *User Datagram Protocol* (UDP).

Characteristics

The creators of CoAP explicitly designed it to resemble HTTP. Like HTTP, it is a request/response protocol relying on *Uniform Resource Locators* (URLs), content types, and options. You can use proxies and caching for CoAP traffic like for HTTP. CoAP requests rely on the same GET, PUT, POST, and DELETE methods as HTTP, although there are subtle differences in the semantics involved. The relationship between CoAP and HTTP is characterized like this in RFC 7252:

> *The goal of CoAP is not to blindly compress HTTP [RFC2616], but rather to realize a subset of REST common with HTTP but optimized for M2M applications. Although CoAP could be used for refashioning simple HTTP interfaces into a more compact protocol, more importantly it also offers features for M2M such as built-in discovery, multicast support, and asynchronous message exchanges.*

Clients send CoAP requests to servers to execute an action specified through a method code on a specific resource. A URI identifies the resource. Servers send back a response described by a response code that may include the representation of a resource. CoAP requests and responses are asynchronous. Since the protocol runs on UDP or a UDP equivalent, which is inherently unreliable, messages can arrive out of order, be duplicated, or even vanish. CoAP implements a lightweight reliability mechanism to address those issues.

Note   Compared to TCP, UDP is a much simpler protocol. It is connectionless and does not perform network handshakes, which means it does not provide ordering, reliability, and data integrity features. This is by design, as UDP delegates the task of implementing such features to the higher layers of the stack. Many IoT protocols leverage UDP as their transport since its simplicity reduces resource requirements. CoAP implements simple reliability and integrity features that can still work on weaker networks that will result in higher message loss.

CoAP's reliability behavior depends on the message type selected. There are currently four message types available:

- **Confirmable:** The message is transmitted reliably. The recipient of a confirmable message must acknowledge it through an acknowledgment message or reject it. If the message is rejected, the recipient will ignore the message and send back a matching reset message.

- **Nonconfirmable:** The message is transmitted unreliably. The recipient must not acknowledge it. If the message is rejected, the recipient may optionally decide to send back a matching reset message.

- **Acknowledgment:** The message is sent in response to a confirmable message. It must echo the original message ID and must contain a response or be empty. There are two types of acknowledgments: piggybacked and empty. Piggybacked acknowledgments carry the response to the request. On the other hand, an empty acknowledgment message only confirms reception of the confirmable message by the recipient, not the success or failure of the request contained in the confirmable message.

- **Reset:** The message is sent in response to a confirmable or nonconfirmable message. A reset message indicates that the recipient received the original message but lacks the context to process it appropriately.

In the absence of an acknowledgment message, the sender of a confirmable message will retransmit the message at an increasing interval until it receives an acknowledgment, or the maximum number of attempts permitted has been reached. CoAP implementations must also detect duplicate messages for both confirmable and nonconfirmable messages.

If you are a developer, when should you pick confirmable messages over nonconfirmable ones? The easiest way to decide is by looking at the value of the commands or data you are sending. If you can afford to miss a reading or have a command not being executed, then nonconfirmable messages are sufficient. Acknowledgments and retransmission are undoubtedly helpful for cases where you wish data and commands eventually reach their destination.

A best practice when working with CoAP – or with most IoT protocols – is to keep message size small. RFC 7959 added support for block-wise transfers to CoAP in 2016, making it possible to use CoAP to transfer larger payloads such as device firmware updates. However, the protocol is better suited to small payloads, and you should explore alternatives if you frequently need to transfer large quantities of data in a limited number of messages.

An essential difference between CoAP and HTTP is that CoAP does not use headers but rather message options. Those options apply to both requests and responses. For example, the URI of a request is transmitted in several options, making parsing more efficient. Section 5.4 of RFC 7252 lists the options available. Another aspect to consider is that message options are binary rather than text-like HTTP headers, significantly reducing message size.

One interesting option was added in RFC 7641, titled *"Observing Resources in the Constrained Application Protocol (CoAP)."* The *observe* option is a widely supported extension to CoAP that enables requesters to retrieve a resource representation and keep it updated over a defined period. One could say CoAP observes comparable to subscriptions, although RFC 7641 explicitly states that the feature's intent is not to replace publish/subscribe networks.

Finally, it is worth mentioning that CoAP enables requesters to discover the resources exposed by a specific server. This is achieved by sending a request to the `/.well-known/core` URI, which will respond with a list of the resources accompanied by their respective media types. To discover the servers available on all the reachable networks, you could send a nonconfirmable request to the relevant IP multicast group. For IPv4, this is the `224.0.1.187` address. For IPv6, this is the `FF0X::FD` address; servers should listen in the Link-Local and Site-Local scopes only. Please note that each server can choose to listen for those multicast requests or not.

Caution   Many networks will filter out or block multicast traffic since it can easily be used to mount *denial-of-service* attacks. Before relying on multicast requests for discovery, please make sure to check with your friendly neighborhood network administrator that they will allow them.

Protocol Stack

CoAP is tied to the IP networking protocol. Figure 2-1 illustrates the protocol stack for CoAP, which I will now discuss.

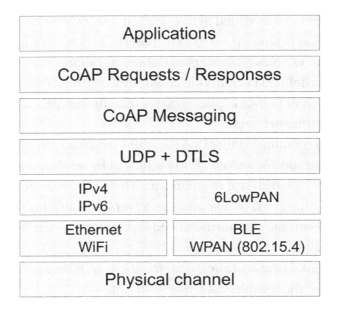

Figure 2-1. *The CoAP protocol stack*

On traditional wired (Ethernet) and wireless (Wi-Fi) networks, CoAP supports both IPv4 and IPv6. However, it is easier to use on the top of IPv6 since all devices are directly routable; using IPv4 behind a *Network Address Translation* (NAT) proxy is always more complex. Whatever IP version is used is irrelevant to the other layers of the stack and, apart from the format of the URIs, does not make a difference in the code you write.

CoAP can also run over *6LoWPAN*, a binding for IPv6 over *Wireless Personal Area Networks* (WPAN). 6LoWPAN allows IPv6 packets to be sent and received over networks based on the IEEE 802.15.4 standard – which is also the basis for Zigbee, Thread, and Snap – or over *Bluetooth Low Energy* (BLE), which has many similar characteristics. The Linux kernel already supports 6LoWPAN, which means you can leverage CoAP over 6LoWPAN on most single-board computers equipped with a Bluetooth radio.

The *Internet Assigned Numbers Authority* (IANA) assigned the following ports and service names to the protocol:

- Port number 5683 and service name "coap" to regular traffic

- Port number 5684 and service name "coaps" to DTLS-secured traffic

Security

Most of the security measures in CoAP rely on DTLS. DTLS is designed to provide security features to datagram protocols such as UDP. Its creators deliberately made it as close to TLS as possible to benefit from the latter's extensive track record and maximize the reuse of existing infrastructure and code. RFC 7252 states it plainly: *"In practice, DTLS is TLS with added features to deal with the unreliable nature of the UDP transport."*

CoAP offers a mechanism to provide configuration values to a new device, including the security information the device needs, such as encryption keys and access control lists. Once provisioning is complete, the device operates in one of four security modes. Those are the following:

- **NoSec:** Security is disabled (DTLS is not used). It is still possible to secure traffic by leveraging lower-layer security; IPsec is an option. I strongly encourage you not to use this mode, even for prototyping. Setting up proper security from the get-go is a better approach.

- **PreSharedKey:** DTLS is enabled, and a list of pre-shared keys is available. Each key specifies a list of nodes it applies to. This mode is the easiest to implement but will not scale well as you add devices.

- **RawPublicKey:** DTLS is enabled, and the device possesses a raw public key – an asymmetric key pair without a certificate. The device has a computed identity derived from the public key in its possession. It also maintains a list of the identities of the devices it is allowed to communicate with. This mode works well with *Public Key Infrastructure* (PKI) and avoids the complexities related to certificates.

- **Certificate:** DTLS is enabled, and the device possesses an asymmetric key pair accompanied by an X.509 digital certificate bound to its subjects and signed by a trust root. The device also has a list of trust root anchors it can contact to validate certificates. Each device should have a unique ID used in the certificate's subject; relying on the IP address would be a bad idea since those are susceptible to change over time. Certificate mode entails more complexity and will require your organization to manage its own PKI. However, if you are building enterprise IoT solutions, you should aim for it if possible.

RFC 7252 states that only the NoSec and RawPublicKey modes are mandatory to implement. Carefully check the documentation of whatever implementation you plan to use before making any security-related decisions.

When DTLS is enabled, the DTLS session and epoch must be the same when matching CoAP responses to a request. The same is true when matching an acknowledgment message or reset message to a confirmable message or a reset message to a nonconfirmable message.

The OSCORE extension (RFC 8613) represents an additional way to secure your CoAP messages. OSCORE protects the messages between the endpoints. Specifically, it protects the request method, the resource identifier, the content type, and the payload of the messages. To achieve this, OSCORE leverages the *Concise Binary Object Representation* (CBOR) format for encoding and *CBOR Object Signing and Encryption* (COSE) for encryption. OSCORE's overhead can be as low as 11 bytes over standard CoAP. OSCORE applies protection on a field-by-field basis. The values of protected fields are encrypted, and their integrity is guaranteed. Proxies can still access unprotected fields while the messages are in transit.

Eclipse Californium

Eclipse Californium is a widely used and mature CoAP implementation written in Java. The project has been active since April 2014 and is used by several other Eclipse IoT projects. Californium is available under two licenses: the Eclipse Distribution License v1.0 (similar to the 3-Clause BSD license) and the Eclipse Public License v2.0.

The official web resources for Californium are as follows:

- **Website:** https://eclipse.org/californium

- **Eclipse project page:** https://projects.eclipse.org/projects/iot.californium

- **Code repository:** https://github.com/eclipse/californium

- **Wiki:** https://github.com/eclipse/californium/wiki

As of writing, the most recent version of Californium was v3.0.0, released on November 3, 2021. Californium integrates its own implementation of DTLS v1.2. Named Scandium, it is available in the principal Californium repository.

Californium v.3.0.0 offers the following stable features:

- CoAP (RFC 7252)

- Observe/notify (RFC 7641)

- Block-wise transfers (RFC 7959)

- No Server Response (RFC 7967)

- Resource Directory draft (draft-ietf-core-resource-directory-20) implementation

- DTLS 1.2 (RFC 6347) through Scandium

- DTLS 1.2 Connection ID (draft-ietf-tls-dtls-connection-id-13)

- DTLS Record Size Limit Extension (RFC 8449)

- DTLS Extended Master Secret (RFC 7627)

It also includes a few experimental features that could become stable in future versions:

- RSA in DTLS support

- Bouncy Castle in DTLS (alternative to OpenJDK JCE) support

- CoAP over TCP (RFC 8323) support – incomplete

- OSCORE support

Let's now see if you can build your CoAP application using the Californium framework. To do this, I will rely on snippets taken from the numerous demo apps shipping with Californium itself.

Sandbox Server

A public sandbox server is available at `coap://californium.eclipseprojects.io:5683/`. It relies on the `cf-plugtest-server` demo application and always uses the most recent code committed to the master branch.

Getting Started

Since Californium is written in Java, it uses *Apache Maven* as its build system. Assuming you have a working Maven installation available and a JDK in your path, you can create an executable JAR by cloning the GitHub repository and running `mvn clean install -DskipTests` from the root folder. Doing so will also give you JARs for the various demo applications in the `demo-apps/run` subfolder.

If you plan to use Maven for your project, you need to add a dependency to the relevant `pom.xml` file as shown in Listing 2-1.

Listing 2-1. Maven dependency declaration for Californium

```
<dependencies>
  ...
  <dependency>
        <groupId>org.eclipse.californium</groupId>
        <artifactId>californium-core</artifactId>
        <version>3.0.0</version>
  </dependency>
  ...
</dependencies>
...
```

Simple GET Request: Demo

The `cf-helloworld-client` demo illustrates how to send a request to a CoAP server. If you cloned the repository and built Californium, you can execute the demo client by issuing the following command:

```
java -jar cf-helloworld-client-3.0.0.jar GETClient coap://californium.
eclipseprojects.io:5683/
```

In this case, I send the request to the public sandbox. If there are no issues, you will get output resembling this:

```
10:21:07.463 INFO [Configuration]: defaults added COAP.
10:21:07.466 INFO [Configuration]: defaults added SYS.
10:21:07.466 INFO [Configuration]: defaults added UDP.
```

```
10:21:07.468 INFO [Configuration]: loading properties from file
...
10:21:07.699 INFO [CoapEndpoint]: coap Started endpoint at coap://
[0:0:0:0:0:0:0:0]:39470
10:21:07.699 INFO [EndpointManager]: created implicit endpoint coap://
[0:0:0:0:0:0:0:0]:39470 for coap
2.05
{"Content-Format":"text/plain", "Size2":457}
****************************************************************
CoAP RFC 7252                               Cf 3.1.0-SNAPSHOT
****************************************************************
This server is using the Eclipse Californium (Cf) CoAP framework
published under EPL+EDL: http://www.eclipse.org/californium/

(c) 2014-2021 Institute for Pervasive Computing, ETH Zurich and others
****************************************************************
```

You can see the server response at the bottom.

Simple GET Request: The Code

Let's now have a look at the main parts of the code for the simple GET request demo.

By default, Californium stores various configuration parameters in a Java .properties file. It is possible to override those default values, however. The following code snippet, which reproduces the top part of the class containing the simple GET request demo, shows how to accomplish this:

```
private static final File CONFIG_FILE = new File("Californium3.properties");
private static final String CONFIG_HEADER = "Californium CoAP Properties
file for client";
private static final int DEFAULT_MAX_RESOURCE_SIZE = 2 * 1024 *
1024; // 2 MB
private static final int DEFAULT_BLOCK_SIZE = 512;

static {
    CoapConfig.register();
    UdpConfig.register();
}
```

```
private static DefinitionsProvider DEFAULTS = new DefinitionsProvider() {

    @Override
    public void applyDefinitions(Configuration config) {
        config.set(CoapConfig.MAX_RESOURCE_BODY_SIZE, DEFAULT_MAX_
        RESOURCE_SIZE);
        config.set(CoapConfig.MAX_MESSAGE_SIZE, DEFAULT_BLOCK_SIZE);
        config.set(CoapConfig.PREFERRED_BLOCK_SIZE, DEFAULT_BLOCK_SIZE);
    }
};
```

You can see that the DefinitionsProvider static inner class applies specific default values for some of the configuration parameters; Californium will assign default values by itself for every parameter not specified by the user.

Before doing anything else, the main (String args[]) method initializes the configuration. createWithFile will create a .properties file if none exists or read from an existing file if one is there.

```
Configuration config = Configuration.createWithFile(CONFIG_FILE,
CONFIG_HEADER, DEFAULTS);
Configuration.setStandard(config);
```

Once we have a valid configuration, sending the GET request to the URI specified on the command line is straightforward. The response variable of type CoapResponse gives us access to the contents of the server response. The following snippet shows how to send the GET request and process the response:

```
CoapClient client = new CoapClient(uri);

try {
    CoapResponse response = client.get();
    if (response != null) {
        System.out.println(response.getCode());
        System.out.println(response.getOptions());
        if (args.length > 1) {
            ...
        } else {
            System.out.println(response.getResponseText());
```

```
            ...
        }
    } else {
        System.out.println("No response received.");
    }
} catch (ConnectorException | IOException e) {
    System.err.println("Got an error: " + e);
}

client.shutdown();
```

Simple Server: Demo

Building a CoAP server is more involved since you need to bind to network interfaces and implement the resources. Fortunately, the Californium demo apps offer several examples of servers. First, let's see how you can run the `cf-helloworld-server`, a simple implementation. To start it, simply issue this command:

```
java -jar cf-helloworld-server-3.0.0.jar HelloWorldServer
```

You should see the following output. Note that since we ran the client earlier, Californium notifies us it is loading the configuration from `Californium3.properties`.

```
11:51:01.585 INFO [Configuration]: defaults added COAP.
11:51:01.588 INFO [Configuration]: defaults added SYS.
11:51:01.588 INFO [Configuration]: defaults added UDP.
11:51:01.589 INFO [Configuration]: defaults added TCP.
11:51:01.590 INFO [Configuration]: loading properties from file
/home/fdesbiens/californium/demo-apps/run/Californium3.properties
11:51:01.626 INFO [RandomTokenGenerator]: using tokens of 8 bytes in length
...
11:51:01.685 INFO [CoapEndpoint]: coap Started endpoint at
coap://127.0.0.1:5683
11:51:01.686 DEBUG [UDPConnector]: Starting network stage thread
[UDP-Sender-/127.0.0.1:5683[1]]
```

By default, this server will start in UDP mode only; you could also enable the experimental TCP support if desired. To verify that the server is running properly, you can use the cf-helloworld-client to send a GET request to the resource it exposes. The name of the resource is helloWorld. Open a new command prompt and execute this command:

```
java -jar cf-helloworld-client-3.0.0.jar GETClient coap://127.0.0.1/
helloWorld
```

You should not encounter networking issues since you are running the client and server from the same machine. There should be a few output lines on the command prompt used to invoke the client. I reproduced the critical part of the output:

```
...
2.05
{"Content-Format":"text/plain"}
Hello World!
...
```

In this case, the server returned me a simple text response. The 2.05 you see is one of the response codes for a successful CoAP GET request – like a 200 in HTTP.

Simple Server: The Code

I will now explore the most important portions of the cf-helloworld-server code with you. For simplicity, the authors kept everything in a single class named HelloWorldServer, which extends Californium's CoapServer class. This class contains the main method used to initialize and start the server. Here is the core of that method:

```
try {
    // create server
    boolean udp = true;
    boolean tcp = false;
    int port = Configuration.getStandard().get(CoapConfig.COAP_PORT);
    ...
    HelloWorldServer server = new HelloWorldServer();
    // add endpoints on all IP addresses
    server.addEndpoints(udp, tcp, port);
    server.start();
```

```
} catch (SocketException e) {
    System.err.println("Failed to initialize server: " + e.getMessage());
}
```

The constructor for the HelloWorldServer class simply instantiates the CoAP resources that the server exposes. You can see it in the following:

```
public HelloWorldServer() throws SocketException {

    // provide an instance of a Hello-World resource
    add(new HelloWorldResource());

    ...
}
```

HelloWorldResource just handles GET requests, as you can see in the next snippet. You would need to override additional methods to handle POST, PUT, and DELETE requests. The base CoapResource class provides default implementations returning response code 4.05 (Method Not Allowed).

```
static class HelloWorldResource extends CoapResource {

    public HelloWorldResource() {

        // set resource identifier
        super("helloWorld");
        // set display name
        getAttributes().setTitle("Hello-World Resource");
    }

    @Override
    public void handleGET(CoapExchange exchange) {

        // respond to the request
        exchange.respond("Hello World!");
    }
}
```

Finally, the addEndpoints method is where CoapEndpoint instances are created for each of the network interfaces available on the device. The whole interaction relies on the well-known *builder* pattern. In this case, the builder is a static nested class defined in CoapEndpoint. The method can also instantiate TCP endpoints if needed. Here is the code for addEndpoints:

```
private void addEndpoints(boolean udp, boolean tcp, int port) {
    Configuration config = Configuration.getStandard();
    for (InetAddress addr : NetworkInterfacesUtil.getNetworkInterfaces()) {
        InetSocketAddress bindToAddress = new InetSocketAddress(addr, port);
        if (udp) {
            CoapEndpoint.Builder builder = new CoapEndpoint.Builder();
            builder.setInetSocketAddress(bindToAddress);
            builder.setConfiguration(config);
            addEndpoint(builder.build());
        }
        if (tcp) {
            TcpServerConnector connector = new TcpServerConnector(bindToAdd
            ress, config);
            CoapEndpoint.Builder builder = new CoapEndpoint.Builder();
            builder.setConnector(connector);
            builder.setConfiguration(config);
            addEndpoint(builder.build());
        }
    }
}
```

About DTLS

The examples I presented up to now use plain CoAP. Fortunately, the Californium team also built a DTLS demo client and server. Looking at the core of the code for the client, you will see that the main difference for sending a GET is the use of DTLSConnector instead of UDPConnector.

```
public class SecureClient {

    ...
    private final DTLSConnector dtlsConnector;
    private final Configuration configuration;
```

```
    public SecureClient(DTLSConnector dtlsConnector, Configuration
    configuration) {
        this.dtlsConnector = dtlsConnector;
        this.configuration = configuration;
    }

    public void test() {
        CoapResponse response = null;
        try {
            URI uri = new URI(SERVER_URI);
            CoapClient client = new CoapClient(uri);
            CoapEndpoint.Builder builder = new CoapEndpoint.Builder().
            setConfiguration(configuration)
                    .setConnector(dtlsConnector);

            client.setEndpoint(builder.build());
            response = client.get();
            client.shutdown();
        } catch (URISyntaxException e) {
            System.err.println("Invalid URI: " + e.getMessage());
            System.exit(-1);
        } catch (ConnectorException | IOException e) {
            System.err.println("Error occurred while sending
            request: " + e);
            System.exit(-1);
        }
    }
}
```

The DTLSConnector instance is initialized in the main method of the SecureClient class. You can see the relevant code as follows:

```
DtlsConnectorConfig.Builder builder = DtlsConnectorConfig.
builder(configuration);
CredentialsUtil.setupCid(args, builder);
...
CredentialsUtil.setupCredentials(builder, CredentialsUtil.CLIENT_NAME, modes);
```

```
DTLSConnector dtlsConnector = new DTLSConnector(builder.build());

SecureClient client = new SecureClient(dtlsConnector, configuration);
client.test();
```

The authors of this demo regrouped logic to handle the security credentials, whether keys or certificates, in the CredentialsUtil class. This class, in turn, relies on the resources found in the demo-certs subfolder of the Californium repository. The README in that folder discusses how the demo keys and certificates are structured and how to leverage the create-keystores.sh script to create your credentials. This script relies on the key and certificate management tools provided by your installed *Java Development Kit* (JDK).

CoAP and Constrained Devices

While Californium is a mature and fully featured CoAP implementation, it cannot be used on constrained devices unable to run high-level operating systems such as Linux or Windows. This is because RTOSes usually lack the required Java runtime. Fortunately, there are libraries available, and some RTOSes even ship with built-in support for CoAP.

In the context of a constrained device, an important design question you will need to answer is whether the device should be a CoAP server or not. Running a server will consume more resources than occasionally sending requests as a client. If all you need to do is report sensor values, then running a server is probably overkill – especially for use cases where you need to maximize battery life. However, if the device needs to execute commands, exposing CoAP resources through a server could make sense. This is the approach chosen by LwM2M, for example.

The Zephyr RTOS from the Linux Foundation ships with a built-in CoAP library; the implementation covers block-wise transfers (RFC 7959) and observe (RFC 7641). Moreover, it supports both client and server roles. The Zephyr CoAP library is implemented using plain buffers and requires developers to manipulate sockets directly. As a reminder, Zephyr is implemented in C.

The Zephyr repository on GitHub contains an extensive collection of samples, including a CoAP client and server. Let's now have a quick look at the client code to see what is involved when sending a GET to a remote server.

First comes the socket declaration.

```
static int sock;
```

The start_coap_client function is called by the main function to set up the socket. Please note that the sample assumes an IPv6 network.

```
static int start_coap_client(void)
{
    int ret = 0;
    struct sockaddr_in6 addr6;

    addr6.sin6_family = AF_INET6;
    addr6.sin6_port = htons(PEER_PORT);
    addr6.sin6_scope_id = 0U;

    inet_pton(AF_INET6, CONFIG_NET_CONFIG_PEER_IPV6_ADDR,
            &addr6.sin6_addr);

    sock = socket(addr6.sin6_family, SOCK_DGRAM, IPPROTO_UDP);
      if (sock < 0) {
            LOG_ERR("Failed to create UDP socket %d", errno);
            return -errno;
      }

    ret = connect(sock, (struct sockaddr *)&addr6, sizeof(addr6));
    if (ret < 0) {
            LOG_ERR("Cannot connect to UDP remote : %d", errno);
            return -errno;
    }

     prepare_fds();

     return 0;
}
```

The actual code to send the request is located in the send_simple_coap_request function, reproduced in a simplified version:

```
static int send_simple_coap_request(uint8_t method)
{
    uint8_t payload[] = "payload";
    struct coap_packet request;
    const char * const *p;
```

```
    uint8_t *data;
    int r;

    data = (uint8_t *)k_malloc(MAX_COAP_MSG_LEN);
    if (!data) {
        return -ENOMEM;
    }

    r = coap_packet_init(&request, data, MAX_COAP_MSG_LEN,
                COAP_VERSION_1, COAP_TYPE_CON,
                COAP_TOKEN_MAX_LEN, coap_next_token(),
                method, coap_next_id());
    if (r < 0) {
        LOG_ERR("Failed to init CoAP message");
        goto end;
    }

    for (p = test_path; p && *p; p++) {
        r = coap_packet_append_option(&request, COAP_OPTION_URI_PATH,
                        *p, strlen(*p));
        if (r < 0) {
            LOG_ERR("Unable add option to request");
            goto end;
        }
    }

    ...
    net_hexdump("Request", request.data, request.offset);

    r = send(sock, request.data, request.offset, 0);
end:
    k_free(data);

    return 0;
}
```

CHAPTER 3

LwM2M

Nous avons besoin les uns des autres. L'être humain n'est pas fait pour s'isoler, mais pour partager.

We need each other. Human beings are not made to isolate themselves, but to share.

—Alice Parizeau, *La charge des sangliers*

Modern applications usually have numerous dependencies. As developers, you and I are standing on the shoulders of giants. Or rather, our own code stands on the top of the many libraries we use. Many of the technologies I cover in this book are built on previous advances. LwM2M, the focus of this chapter, is a great example of that.

Historically, LwM2M exclusively used CoAP as a transport. While version 1.2 of LwM2M added support for HTTP and MQTT transports at the end of 2020, CoAP is still the most prevalent one used. Therefore, my code examples will leverage Eclipse Leshan, a framework to build LwM2M solutions that uses Californium as its CoAP implementation.

LwM2M: Built on CoAP's Foundation

CoAP provides a solid foundation to build IoT solutions. In particular, the fact that message payloads are not defined by the various RFCs describing the protocol means that CoAP is very flexible. The same flexibility applies to resources; developers can structure them in any way they see fit. The consequence is that, out of the box, devices and software stacks that support CoAP are not interoperable. At a minimum, you will need to massage payloads and adjust some parameters to ensure that messages are sent to the right resource in the right format.

© Frédéric Desbiens 2023
F. Desbiens, *Building Enterprise IoT Solutions with Eclipse IoT Technologies*,
https://doi.org/10.1007/978-1-4842-8882-5_3

The *Lightweight Machine-to-Machine* (LwM2M) protocol aims to solve this lack of interoperability by offering an extensible resource and data model. It is under the governance of OMA SpecWorks, previously the Open Mobile Alliance (OMA), a nonprofit standards organization. Version 1.0 of the protocol was published in February 2017, and the latest version at the time of writing was version 1.2, published in November 2020.

Data Model: Objects and Resources

The LwM2M data model is based on *resources*: elements of information made available by devices. Resources are logically grouped into *objects*. Each resource is assigned a unique identifier within the enclosing object. OMA SpecWorks assigns and maintains a unique *Object Identifier* for all objects that are a core part of the LwM2M specification and object specifications. Object specifications define the operations (Read, Write, Execute) supported by the object's resources and whether the resources are mandatory or optional. They also state the data type of each resource and whether there can be multiple instances or not.

Third parties such as standards organizations, individuals, or vendors can propose object specifications to OMA SpecWorks. When accepted, such object specifications are made available in the official LwM2M registry and allocated a unique ID. For example, the LwM2M specification defines a Device object (ID: 3) used to expose device information to interested parties. Device contains resources for the name of the device manufacturer, model number, and serial number, among other things. Moreover, the registry also contains reusable resource definitions that you can use in your Object specifications. Such resource definitions are also allocated unique IDs for reference purposes. Typical examples include digital input, digital output, analog input, analog output, dimmer value, and unit of measurement. Sensor Value (ID: 5700) is one of the reusable resources defined in the registry and is used in many objects reporting sensor readings. One of those objects is Temperature (ID: 3303). It is worth mentioning that object specifications are versioned, whether resource definitions are not.

Note   The official LwM2M registry is available at

```
https://technical.openmobilealliance.org/OMNA/LwM2M/
LwM2MRegistry.html
```

At runtime, clients and servers will instantiate objects and their resources. An object can contain multiple instances of the same resource, and multiple instances of the same object can exist simultaneously. Objects and the resource they contain are referenced by their assigned numerical ID, while object and resource instances get an instance number assigned at runtime. Instance numbers start at 0 – as they should be. All IDs and instance numbers are unsigned integers.

The LwM2M resource model includes operations for creating, updating, and retrieving resources and enables asynchronous notifications of resources changes. There are three operations a resource can support: Read, Write, and Execute. Resources that support Read or Write are called *value* resources. Resources that support Execute are called *executables*; they are used to trigger actions and do not contain a value.

To access a resource, all you need to do is refer to it using a simple URI. This is the pattern you should use:

```
/[object id]/[object instance]/[resource id]/[resource instance]
```

All of this may sound a little abstract. So let's review a concrete example. The LwM2M specification defines an object called `Device` (ID: 3), which defines a set of resources providing device information and executables to reboot the device or perform a factory reset of the settings. Table 3-1 lists a few of the resources for the `Device` object in version v1.1 of LwM2M.

Table 3-1. *Partial list of resources for the Device LwM2M object*

ID	Name	Operations	Instances	Mandatory	Type
0	Manufacturer	Read	Single	Optional	String
1	Model Number	Read	Single	Optional	String
2	Serial Number	Read	Single	Optional	String
3	Firmware Version	Read	Single	Optional	String
4	Reboot	Execute	Single	Mandatory	
5	Factory Reset	Execute	Single	Optional	

The LwM2M specification states that devices must implement the `Device` object and that only a single instance will exist. Consequently, any device will have an instance of the `Device` object exposed as

/3/0

If the device in question implements the `Manufacturer` resource, it will be available as

/3/0/0

Let's look at another example. The LwM2M registry contains a `Temperature` object used to report values from a temperature sensor. The ID of this object is 3303. This object declares several resources, of which one is `Sensor Value`. `Sensor Value` is one of the reusable resources defined in the registry, and its ID is 5700. So to retrieve the current temperature reported by a device equipped with a single temperature sensor, you would query the resource in this way:

/3303/0/5700

LwM2M can leverage several serialization formats for data: Plain Text, Opaque, TLV, LwM2M TS 1.0 JSON, CBOR, LwM2M CBOR, SenML JSON, and SenML CBOR. Binary formats such as CBOR are more efficient since they compress the size of the payloads. The specification states that supporting all the formats is mandatory for servers.

Additional Capabilities over CoAP

CoAP's RESTful model heavily inspires LwM2M's messaging model. The additional interfaces LwM2M defines between devices and servers make it different – and arguably, better. Those are the following:

- Bootstrap
- Client Registration
- Device Management and Service Enablement
- Information Reporting

I will now cover each of these in detail.

Bootstrap

The bootstrap interface is used to provision configuration parameters in a client to enable it to perform client registration against one or several LwM2M servers.

There are four distinct bootstrap modes:

- **Factory Bootstrap:** The device has been configured before deployment. The configured information can be related to a LwM2M bootstrap server or LwM2M servers.

- **Bootstrap from Smartcard:** The device reads configuration information from a smart card. This supposes it possesses the hardware required to read such a card. Ideally, the information is read through a secure channel, and the device will validate that it matches whatever is stored on the smart card.

- **Client Initiated Bootstrap:** The device will retrieve configuration information from a LwM2M bootstrap server. This requires a LwM2M bootstrap server account to be present on the client. The client also needs security credentials to connect to the bootstrap server, whether of the TLS/DTLS or OSCORE variety.

- **Server Initiated Bootstrap:** An authorized LwM2M server triggers a bootstrap sequence in the device. The device will then switch modes to perform a client-initiated bootstrap.

Clients are required to support at least one of the modes.

Client Registration

Clients use this interface to register with one or several servers, maintain their registration, and deregister from a server.

When a device registers, it provides a list of the objects it supports and its current object instances. Once registered, the device will update its registrations based on its configuration – usually, a time interval or the security context. Registrations have a lifetime; the server will consider devices that do not send a registration update within the time limit to have deregistered.

When a device shuts down or disconnects, it should perform a deregister operation – although this is not mandatory.

Device Management and Service Enablement

Servers use the Device Management and Service Enablement Interface to access the object instances and resources exposed by registered clients. The interface defines specific operations to that effect: Create, Read, Read-Composite, Write, Write-Composite, Delete, Execute, Write-Attributes, and Discover. Clients will ignore all operations made by a server on this interface until they have completed registration with the server in question.

The normal Read operation can apply to the value of a resource, a resource instance, an array of resource instances, an object instance, or all the instances of a specific object. The Write operation has roughly the same scope, except it focuses on a single object instance. On the other hand, Read-Composite can target any combination of objects, object instances, resources, or resource instances in a single request, whether from the same object or different objects. As for Write-Composite, it can update values of several different resources across distinct instances of one or several objects.

Information Reporting

LwM2M servers use this interface to observe value changes for specific resources exposed by registered clients. The observe can target a single resource or be a composite, targeting a group of resources or a group of resource instances across multiple object instances on the client.

LwM2M Versions

At the time of writing, OMA SpecWorks had published three versions of LwM2M. Given the rapid release cadence, not all implementations support the latest release of the protocol. Consequently, you need to know the version supported by the devices and software you are working with to determine which features will be available.

LwM2M v1.0 (February 2017)

The initial release for LwM2M included the following features:

- Object-based resource model

- Resource operations: Creation, retrieval, update, deletion, and configuration

- Resource observation and notification

- Data serialization formats: TLV, JSON, Plain Text, and Opaque

- UDP and SMS transports

- DTLS-based security

- Queue mode (for sleeping devices)

- Core LwM2M objects: LwM2M Security, LwM2M Server, Access Control, Device, Connectivity Monitoring, Firmware Update, Location, Connectivity Statistics

LwM2M v1.1 (June 2018)

LwM2M added several features to v1.0:

- Improved bootstrapping allowing for incremental upgrades

- Improved support for Public Key Infrastructure (PKI) deployments

- Enhanced registration sequence mechanisms

- Support for LwM2M over TCP/TLS to better support firewalls and NAT

- Support for OSCORE-based application layer security

- Improved support for low-power WANs, notably 3GPP CIoT and LoRaWAN

- Support for JSON using SenML with CBOR serialization

- New data types

LwM2M v1.2 (November 2020)

LwM2M is backward-compatible with v1.0 and v1.1 at the level of mandatory features. It introduced the following new features:

- Support for MQTT and HTTP as transports.

- Optimized bootstrapping, registration, and information interfaces.

- Support for LwM2M gateways, allowing the integration and management of non-LwM2M devices.

- New optimized serialization format based on CBOR: LwM2M CBOR.

- Enhanced functionality for firmware updates.

- Definition of new notification attributes (edge, confirmable notification, and maximum historical queue).

- TLS and DTLS v1.3 are now supported.

LwM2M Protocol Stack

The LwM2M protocol stack varies according to the version considered. I will focus on the stack for version 1.2 here since it is the most comprehensive.

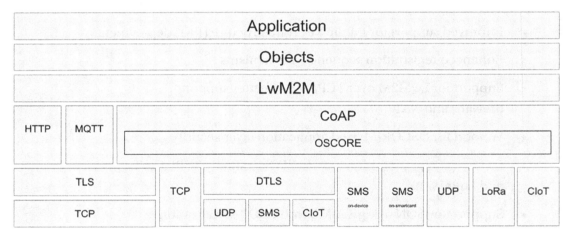

Figure 3-1. *LwM2M 1.2 protocol stack (Credit: OMA SpecWorks)*

In versions before 1.2, LwM2M systematically relied on CoAP as its transport. However, LwM2M always supported a choice of low-level transports to run CoAP on. Initially, the selection was limited to UDP and *Short Message Service* (SMS) over cellular networks. Support for TCP was added in version 1.1 to offer better integration with firewalls and to support NAT traversal scenarios. Version 1.1 also introduced options for CoAP over non-IP protocols, namely, *Cellular Internet of Things* (CIoT) from the *3rd Generation Partnership Project* (3GPP) and *LoRaWAN*. Developers can leverage the OSCORE security model in version 1.1 and up of LwM2M, independently of the low-level transport selected.

Version 1.2 decoupled LwM2M from CoAP, introducing HTTP and MQTT as alternate transports. In all versions, security is based on DTLS or TLS, depending on the transport used.

Eclipse Leshan

The Eclipse Leshan project provides Java libraries that you can use to build your own LwM2M clients and servers or to add LwM2M support to existing platforms. The project has existed since 2014 and uses Eclipse Californium as its CoAP implementation. Leshan is made available under the Eclipse Public License v2.0 and the Eclipse Distribution License v1.0 (BSD).

The official web resources for Leshan are as follows:

- **Website:** https://eclipse.org/leshan

- **Eclipse project page:** https://projects.eclipse.org/projects/iot.leshan

- **Code repository:** https://github.com/eclipse/leshan

- **Wiki:** https://github.com/eclipse/leshan/wiki

The Leshan project team also maintains a demo client, server, and bootstrap server. Those are valuable examples of utilizing the Leshan API and can also be used for troubleshooting or testing purposes.

At the time of writing, Leshan had two main versions under active maintenance:

- **Version 1.x.** These series implement v1.0.2 of the LwM2M specification.

- **Version 2.x.** These series implement v1.1.1 of the LwM2M specification.

The Leshan wiki describes for each version which LwM2M features they support.

Sandbox Server

The team of the Leshan project maintains a pair of publicly available servers that you can use for demo and testing purposes. Those servers always run the last successful build from the master branch.

- **LwM2M server:** https://leshan.eclipseprojects.io/. Use port 5683 for coap and 5684 for coaps.

- **Bootstrap server:** https://leshan.eclipseprojects.io/bs/. Use port 5783 for coap and port 5784 for coaps.

If you would rather use C than Java for building your LwM2M solution, Eclipse Wakaama is a C implementation of LwM2M designed to be portable on POSIX-compliant systems.

Quick Test Drive

The easiest way to test drive Leshan is to run the demo client and server instances on your local machine. To do so, all you need is a JDK available in your path; Leshan requires Java 8 or later.

To download and start the latest binary of the Leshan demo server on Linux, simply run the following two commands:

```
wget https://ci.eclipse.org/leshan/job/leshan/lastSuccessfulBuild/artifact/
leshan-server-demo.jar
java -jar ./leshan-server-demo.jar
```

This will start the server in default mode; it will listen to all available interfaces over IPv4 and IPv6 for coap and coaps traffic on ports 5683 and 5684, respectively.

To download and start the client on Linux, issue the following commands on another command prompt:

```
wget https://ci.eclipse.org/leshan/job/leshan/lastSuccessfulBuild/artifact/
leshan-client-demo.jar
java -jar ./leshan-client-demo.jar
```

Once this is done, you can access the server's graphical interface by connecting to `http://localhost:8080` in your browser. You should see a page like the one shown in Figure 3-2.

Figure 3-2. *Eclipse Leshan test server – list of devices*

By default, the client will take the hostname of the machine it runs on. In this case, the name of my trusty workstation is `Ashitaka`.

Note   Ashitaka is the protagonist of *Princess Mononoke* by Hayao Miyazaki. You should watch it!

If you click on the device, you will access a page giving access to its objects and resources. Figure 3-3 shows the list of attributes for the `Device` object.

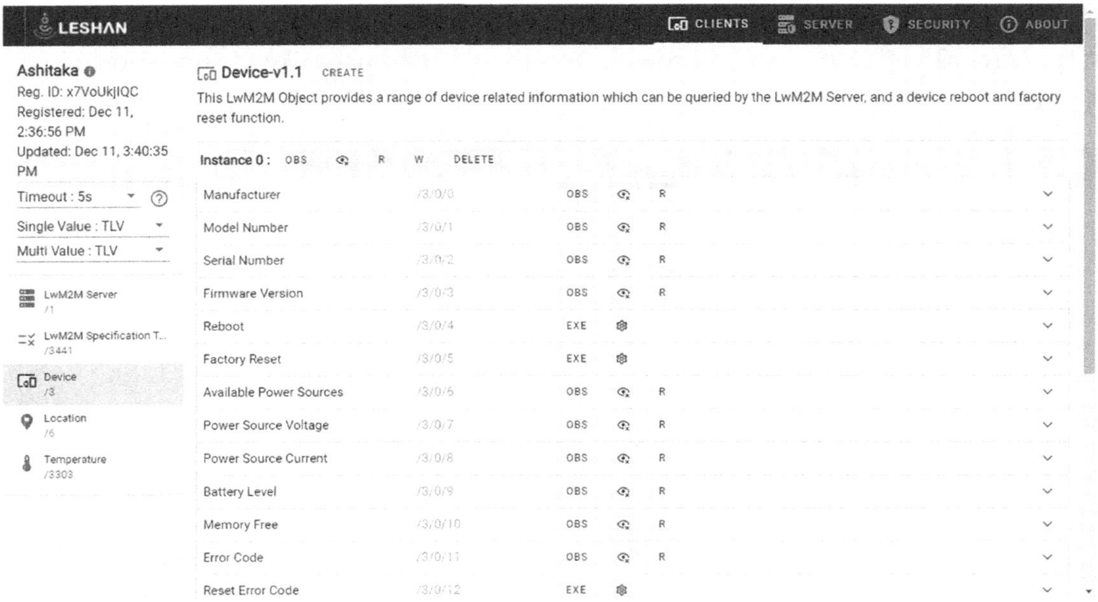

Figure 3-3. *Eclipse Leshan test server – list of resources, Device object*

If you click on the arrow at the end of a line for a resource, you will see detailed information about the resource in question. In Figure 3-4, I accessed the details of the Manufacturer resource. It is a read-only singleton resource of type String. The OBS and "eye" buttons can start and stop an observe on it. The R button reads the value, which is "Leshan Demo Device." There is a circle on the button to indicate I clicked it already.

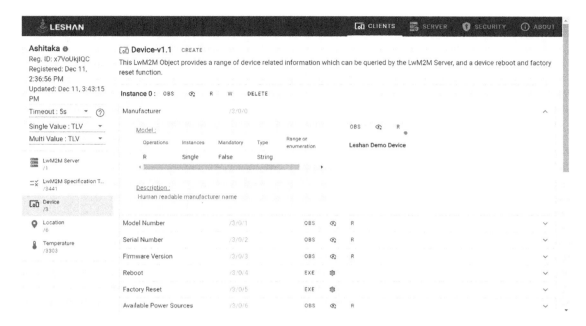

Figure 3-4. *Eclipse Leshan test server – resource value and details*

The Leshan demo server's UI makes it very easy to access various information about the server, including the URLs for its endpoints, its public key (for raw public key mode), and its X.509 server certificate. Figure 3-5 shows how that page looks.

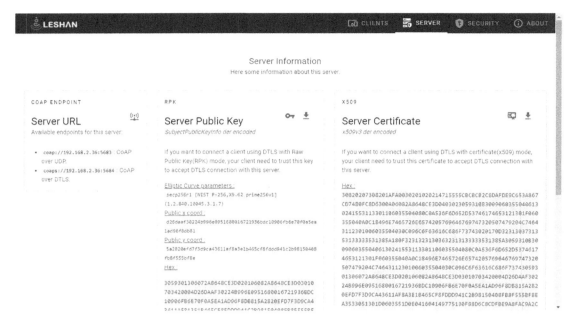

Figure 3-5. *Eclipse Leshan test server – server information*

Building Your Client

Like Californium, Leshan relies on the Maven build system. To start building your own client, you will need to add the dependencies shown in Listing 3-1 to your pom.xml:

Listing 3-1. Maven dependency declaration for a Leshan client

```
<dependencies>
    <dependency>
        <groupId>org.eclipse.leshan</groupId>
        <artifactId>leshan-client-cf</artifactId>
        <version><!-- use the lastest version --></version>
    </dependency>
    <!-- add any slf4j backend -->
    <dependency>
        <groupId>org.slf4j</groupId>
        <artifactId>slf4j-simple</artifactId>
        <version>1.7.30</version>
        <scope>runtime</scope>
```

```
    </dependency>
</dependencies>
```

To create a Leshan client instance, you need to leverage the `LeshanClientBuilder` class. After that, all you need to do is call the `client` object's `start` method. By default, Leshan will connect to the public demo server at leshan.eclipseprojects.io if no other parameters are provided. The following code sample shows how to create and start the client:

```
String endpoint = "..." ; // choose an endpoint name
LeshanClientBuilder builder = new LeshanClientBuilder(endpoint);
LeshanClient client = builder.build();
client.start();
```

LeshanClientBuilder creates instances of the `Security`, `Server`, and `Device` objects populated with default values. Let's say we wish to override the default values for those three and implement another object, `ConnectivityStatistics`, ourselves. The code for ConnectivityStatistics looks like this:

```
public class ConnectivityStatistics extends BaseInstanceEnabler {

    @Override
    public ReadResponse read(ServerIdentity identity, int resourceid) {
        switch (resourceid) {
        case 0:
            return ReadResponse.success(resourceid, getSmsTxCounter());
        }
        return ReadResponse.notFound();
    }

    @Override
    public WriteResponse write(ServerIdentity identity, int resourceid,
    LwM2mResource value) {
        switch (resourceid) {
        case 15:
            setCollectionPeriod((Long) value.getValue());
            return WriteResponse.success();
        }
```

```
            return WriteResponse.notFound();
    }

    @Override
    public ExecuteResponse execute(ServerIdentity identity, int resourceid,
    String params) {
        switch (resourceid) {
        case 12:
            start();
            return ExecuteResponse.success();
        }
        return ExecuteResponse.notFound();
    }
}
```

Please note that all the numerical identifiers in the code refer to the LwM2M registry.

Leshan provides the ObjectsInitializer class to create instances of LwM2M objects and implementations of Security, Server, and Device. Before calling build to obtain our client instance, we need to call setObjects on the builder (LeshanClientBuilder). The following snippet illustrates how to leverage ObjectsInitializer:

```
ObjectsInitializer initializer = new ObjectsInitializer();
initializer.setInstancesForObject(LwM2mId.SECURITY, Security.noSec("coap://
localhost:5683", 12345));
initializer.setInstancesForObject(LwM2mId.SERVER, new Server(12345, 5 * 60,
BindingMode.U, false));
initializer.setInstancesForObject(LwM2mId.DEVICE, new Device("Ghibli
Savoia", "S.21", "12345", "U"));
initializer.setInstancesForObject(7, new ConnectivityStatistics());

// add it to the client
builder.setObjects(initializer.createAll());
LeshanClient client = builder.build();
client.start();
```

You should study the code of the Leshan client demo for a complete reference.

Building Your Own Server

Building a basic LwM2M server with Leshan is even easier than for a client. The Maven dependencies look the same, but the `artifactId` is different. The client and server libraries are distinct.

Listing 3-2. Maven dependency declaration for a Leshan server

```
<dependencies>
    <dependency>
        <groupId>org.eclipse.leshan</groupId>
        <artifactId>leshan-server-cf</artifactId>
        <version><!-- use the lastest version --></version>
    </dependency>
    <!-- add any slf4j backend -->
    <dependency>
        <groupId>org.slf4j</groupId>
        <artifactId>slf4j-simple</artifactId>
        <version>1.7.30</version>
        <scope>runtime</scope>
    </dependency>
</dependencies>
```

Creating a basic server with default settings requires only three lines of code:

```
LeshanServerBuilder builder = new LeshanServerBuilder();
LeshanServer server = builder.build();
server.start();
```

This server will accept registrations but will not do much else.

Key Java objects in the Leshan API allow you to use listeners to handle specific events. This is the case of classes implementing the `RegistrationService` interface involved in device registration. Here is how you could create listeners that will fire when a device registers, updates its registration, and unregisters:

```
Server.getRegistrationService().addListener(new RegistrationListener() {
```

```
    public void registered(Registration registration, Registration
    previousReg,
            Collection<Observation> previousObsersations) {
        System.out.println("new device: " + registration.getEndpoint());
    }

    public void updated(RegistrationUpdate update, Registration updatedReg,
    Registration previousReg) {
        System.out.println("device is still here: " + updatedReg.
        getEndpoint());
    }

    public void unregistered(Registration registration,
    Collection<Observation> observations, boolean expired,
            Registration newReg) {
        System.out.println("device left: " + registration.getEndpoint());
    }
});
```

It is possible to send a request to the device when the listener is executing. Here is code you could append to the `registered` method in the preceding snippet to retrieve the value for the current UNIX time of the client device (`Current Time` resource) in its Device object. The URI for the resource is 3/0/13.

```
Try {
        ReadResponse response = server.send(registration, new
        ReadRequest(3,0,13));
        if (response.isSuccess()) {
            System.out.println("Device time:" + ((LwM2mResource)response.
            getContent()).getValue());
        }else {
            System.out.println("Failed to read:" + response.getCode() + " "
            + response.getErrorMessage());
        }
    } catch (InterruptedException e) {
        e.printStackTrace();
    }
```

You could improve the preceding example by providing a response callback and an error callback as additional parameters to the send method. If you do so, the request would be sent asynchronously, which avoids timeouts and other issues.

The Leshan server demo is a useful resource to learn about the Leshan API. In particular, the LeshanServerDemo class shows how to create a DTLS configuration and load the keys and certificates required by the raw public key and X.509 security modes.

LwM2M and Constrained Devices

When using LwM2M, constrained devices are considered clients. Given the bandwidth and processing power requirements, a LwM2M server only makes sense if deployed on a gateway or edge server.

Since the Zephyr RTOS supports CoAP, it is not surprising that it also ships with a LwM2M client library. The library implements v1.0.2 of LwM2M and possesses the following features:

- Engine to process networking events and core functions

- Resource Directory (RD) client, which performs BOOTSTRAP and REGISTER functions

- Support for the TLV, JSON, and Plain Text serialization formats

- Implementations of core LwM2M objects such as Security, Server, Device, and Firmware Update

- Implementations of extended IPSO objects such as Light Control, Temperature Sensor, and Timer

The Zephyr team wrote a sample application showcasing most of the library's features. I will now discuss its most relevant parts.

The main function, reproduced as follows, calls lwm2m_setup before anything else. It will then start the LwM2M client by passing some hardware-related information to lwm2m_rd_client_start.

```
Static struct lwm2m_ctx client;
void main(void)
{
    ret = lwm2m_setup();
```

```
    (void)memset(dev_id, 0x0, sizeof(dev_id));

    length = hwinfo_get_device_id(dev_id, sizeof(dev_id));
    for (i = 0 ; i < length ; i++) {
        sprintf(&dev_str[i*2], "%02x", dev_id[i]);
    }
    lwm2m_rd_client_start(&client, dev_str, rd_client_event);

    k_sem_take(&quit_lock, K_FOREVER);
}
```

The role of the lwm2m_setup function is to declare the objects exposed by the device and their values. Resource values are retrieved from constants declared at the top of the file in the following snippet. Executable resources are assigned callback references.

```
static int lwm2m_setup(void)
{
    ...
    /* setup DEVICE object */
    lwm2m_engine_set_res_data("3/0/0", CLIENT_MANUFACTURER,
    sizeof(CLIENT_MANUFACTURER), LWM2M_RES_DATA_FLAG_RO);
    lwm2m_engine_set_res_data("3/0/1", CLIENT_MODEL_NUMBER,
    sizeof(CLIENT_MODEL_NUMBER), LWM2M_RES_DATA_FLAG_RO);
    lwm2m_engine_set_res_data("3/0/2", CLIENT_SERIAL_NUMBER,
    sizeof(CLIENT_SERIAL_NUMBER), LWM2M_RES_DATA_FLAG_RO);
    lwm2m_engine_set_res_data("3/0/3", CLIENT_FIRMWARE_VER,
    sizeof(CLIENT_FIRMWARE_VER), LWM2M_RES_DATA_FLAG_RO);
    lwm2m_engine_register_exec_callback("3/0/4", device_reboot_cb);
    lwm2m_engine_register_exec_callback("3/0/5", device_factory_
    default_cb);
    ...

    return 0;
}
```

One of the parameters passed to lwm2m_rd_client_start is a reference to rd_client_event. This function is where you could handle specific events. The current implementation simply logs error messages, as you can see in the following listing.

```
static void rd_client_event(struct lwm2m_ctx *client,
                enum lwm2m_rd_client_event client_event)
{
    switch (client_event) {
    [...]
    case LWM2M_RD_CLIENT_EVENT_REGISTRATION_FAILURE:
        LOG_DBG("Registration failure!");
        break;

    case LWM2M_RD_CLIENT_EVENT_REGISTRATION_COMPLETE:
        LOG_DBG("Registration complete");
        break;
    [...]
    }
}
```

Working with 6LoWPAN

If you deploy the Zephyr LwM2M sample on a Bluetooth-equipped device, you could try to establish a 6LoWPAN connection between the constrained device and the LwM2M server. To do so, you need to enable the Bluetooth overlay in your configuration when you flash the constrained device. Then, you need to log in to the Linux machine hosting the LwM2M server and activate the 6LoWPAN kernel module, as shown in the following:

```
sudo su
modprobe bluetooth_6lowpan
echo 1 > /sys/kernel/debug/bluetooth/6lowpan_enable
hcitool lescan
```

The last command will return the list of Bluetooth devices detected by the machine hosting the server. In the output, you need to find the MAC address for the device named LWM2M IPSP node. This name is hard-coded in the sample application. For example:

```
LE Scan ...
C2:FA:8D:93:21:DD LWM2M IPSP node
D8:0F:99:79:2C:DA (unknown)

...
```

Then, all you need to do is establish the connection between that MAC address and the kernel module and assign an IPv6 address to the server's Bluetooth adapter. The Zephyr sample LwM2M application uses a hard-coded address of 2001:db8::1. In this example, the machine's Bluetooth adapter is identified as bt0.

```
echo "connect C2:FA:8D:93:21:DD 2" > /sys/kernel/debug/
bluetooth/6lowpan_control
ip address add 2001:db8::2/64 dev bt0
```

Once this is done, the constrained device should register on the server.

This file provides additional information and describes how to enable DTLS in the Zephyr sample application. I will let you guess which LwM2M server the documentation's writers use.

CHAPTER 4

MQTT

Personne ne sait comment sont exactement les choses quand on ne les regarde pas.

No one knows how things exactly are when you don't look at them.

—Hubert Reeves, *Patience dans l'azur*

To understand the significance of MQTT, you need to understand the difference between IoT and embedded computing. Microcontrollers and general-purpose microprocessors have been used for a long time in large mechanical or electronic systems; the Apollo guidance computer used to conquer the moon in the 1960s is an early example of the concept. Nowadays, embedded systems power various devices, from small personal ones to larger machines such as industrial robots and assembly lines. They are also at the core of specific subsystems of complex machines, such as the antilock brakes of your car.

Embedded systems found in factories, water-processing facilities, and electrical grids are nearly always part of a network. This has been the case since the mid-1970s, which saw the emergence of *Supervisory Control and Data Acquisition* (SCADA) solutions. What makes embedded systems different from IoT solutions, then? In Chapter 1, I wrote that IoT devices *"contain embedded hardware and software."* IoT is thus clearly a specific application of embedded technologies. However, embedded systems are often tailor-made for a particular application, and their software is not frequently updated once the manufacturer ships them. On the other hand, IoT devices benefit from a permanent network connection and often gain new features through software updates. In the end, *Internet* connectivity, and not just network connectivity, is what makes IoT different from earlier embedded computing efforts.

In the next two chapters, I will cover two technologies that are at the forefront of the transition from traditional embedded systems to IoT. The first one is MQTT, which was created at the end of the 20th century and has now become the leading protocol for IoT

© Frédéric Desbiens 2023
F. Desbiens, *Building Enterprise IoT Solutions with Eclipse IoT Technologies*,
https://doi.org/10.1007/978-1-4842-8882-5_4

applications. The second is Sparkplug, which opens new possibilities for MQTT users. Both have been instrumental in making industrial processes easier to observe. In other words, they have accelerated the industry's transition from analog to digital technologies.

What Is MQTT?

MQTT is an OASIS Open standard and an ISO recommendation (ISO/IEC 20922). At the time of writing, the newest version is version 5.0, published in March 2019. This chapter refers to version 3.1.1 of the specification unless specifically mentioned.

Note   MQTT originally stood for "MQ Telemetry Transport." The "MQ" in the name referred to the IBM message queuing products that had been the inspiration for MQTT. The MQTT Technical Committee at OASIS Open decided in 2013 to make MQTT the name of the technology and that MQTT should not stand for anything.[1]

The MQTT protocol was invented in 1999 by Andy Stanford-Clark and Arlen Nipper. They needed to monitor oil pipelines via a satellite connection, and existing protocols were not up to the task since they required continuous polling of the sensors. Consequently, they decided to create a new protocol to maximize the battery life of constrained devices deployed in the field while consuming the least bandwidth possible.

MQTT leverages the publish/subscribe interaction model in a centralized architecture. Message publishers and subscribers are *clients*. Those clients connect to one or several *brokers*, receiving published messages and routing them to the appropriate clients. A client thus can be a publisher, a subscriber, or both. In MQTT, publishers and subscribers are completely decoupled from each other. This means several clients could receive a specific message and publish messages to the same destination. Usually, publishers do not know if any subscribers will receive their messages.

Messages

MQTT messages, also called *packets* in the specification, have a simple structure.

---

[1] https://www.oasis-open.org/committees/download.php/49028/OASIS_MQTT_TC_
minutes_25042013.pdf

1. **Fixed header:** Present in all messages, the fixed header is 1 byte in length. The first four bits (bits 7 to 4) represent an unsigned value for the *packet type*. These packet types include PUBLISH, SUBSCRIBE, and CONNECT, among others. The remaining bits (3 to 0) contain *flags* specific to each packet type. As of MQTT 5.0, the specification's authors reserved most flags for future use except in the case of the PUBLISH packet type.

2. **Variable header:** Present in some messages only. The variable header can contain a *packet identifier* and a set of *properties*. This set of properties comprises a length indicator (variable byte integer) and an arbitrary number of properties. Each property consists of an identifier (variable byte integer) and a value. The value's type depends on the specific property being set, and properties can be set in any order.

3. **Payload:** Present in some messages only. Some messages, such as pings (PINGREQ, PINGRESP) and various types of acknowledgments, do not have a payload.

Section 2 of the MQTT specification describes the packet type flag and the values for the variable headers in detail.

Note   A variable byte integer is an encoding of 64-bit unsigned integers, which uses between 1 and 9 bytes. They have the advantage of taking up less space for small values.

A crucial design decision in MQTT was to make the protocol payload agnostic. Specifically, PUBLISH packets contain a payload in byte format. The publisher can transmit anything as a payload; plain text or binary encodings are fine. Consequently, you can use XML, JSON, or any format of your choosing to represent the data. Version 5.0 of MQTT added properties to named Payload Format and Content Type. Payload Format is a byte value. When set to 0, it indicates an "unspecified byte stream"; when set to 1, it means a "UTF-8 encoded payload." If the payload indicator is missing, the broker will assume it is set to 0 to guarantee backward compatibility with prior protocol versions. As for Content Type, it is a UTF-8 (Unicode) encoded string. Applications usually expect a MIME content type, but you can also use arbitrary values.

Topics and Topic Filters

Since MQTT publishers and subscribers are decoupled from one another, the broker needs a way to determine which messages should be made available to each client. To that effect, the broker organizes messages in a hierarchy of *topics*. In MQTT, a topic is simply a UTF-8 string the broker will use to determine which subscribers will get a copy of a specific message. When a client publishes a message, it must specify a topic. On the other hand, subscribers indicate their interest in topics through *topic filters*.

MQTT clients do not need to create topics before publishing or subscribing to one. The broker will accept any valid topic name without preparation or initialization. The maximum length supported for topic names is the maximum size for UTF-8 strings: 65,535 bytes. However, I strongly recommend restraint and keeping your topic names short, as longer names will impact performance and resource utilization.

To establish a topic hierarchy, all you need to do is to use a forward slash ("/" U+002F) to separate the levels in the hierarchy. Here are a few examples of valid topic names:

```
Louvre/1/101/temperature
Sites/USA/California/SanFrancisco/SiliconValley/
e31a57ef-936b-46b8-ac89-3f39c8489f5f/status
vehicles/trucks/647/speed
```

Topic filters can contain wildcard characters. The specification defines two of them. "+" is a single-level wildcard, while "#" is a multilevel wildcard that can only be used at the end of a filter. The best way to understand the difference is through an example. Suppose clients publish sensor readings to the following topics:

```
Louvre/1/101/temperature
Louvre/1/101/humidity
Louvre/1/102/temperature
Louvre/1/102/humidity
Louvre/2/201/temperature
Louvre/2/201/humidity
```

The first number in the topic names represents the floor, and the second is the room number. Suppose a subscriber uses the following topic filter:

```
Louvre/1/+/temperature
```

Then, it will receive the data for the following topics:

```
Louvre/1/101/temperature
Louvre/1/102/temperature
```

It will not get the messages for the second-floor rooms or the humidity sensors. On the other hand, if the filter was like the following:

```
Louvre/2/#
```

then the subscriber would get the messages sent to these topics:

```
Louvre/2/201/temperature
Louvre/2/201/humidity
```

The subscriber would not receive any messages sent under `Louvre/1/` in that case. The MQTT specification defines rules applying to topic names and topic filters:

- They must be at least one character long.

- They are case sensitive.

- They can include the space character.

- They must not contain the null character (Unicode U+0000).

Additionally, topic names must not contain the topic filter wildcard characters ("#" and "+").

Topic names are completely arbitrary. Here are a few recommendations you should follow when picking names for your topics or subscribing:

- **Avoid nonprintable characters:** UTF-8 contains many nonprintable characters, including tabs, spaces, and line breaks. Avoid them since you will not be able to tell them apart. Just because the specification states you can use spaces does not mean it is a good idea.

- **Do not subscribe to "#":** Using just "#" as your topic filter means that the client subscribes to all messages published to a broker. This could overwhelm the client or require significant computing resources in production environments. I strongly recommend pursuing alternative strategies. For example, suppose you need to log every message sent in a historian database. In that case, you could use a broker offering such a feature – which is not part of MQTT itself – or leverage Eclipse Hono and Apache Kafka to deploy a proper stream event processing solution.

- **Build a sensible and extensible topic hierarchy:** The goal of topic hierarchies is to provide context relevant to messages. Consequently, your topics should be as specific as possible. Subscribers should not have to parse the payload to figure out the context of a message. Moreover, your topic hierarchy should accommodate future evolutions. If you add sensors to a machine or simply wish to report about new events, you should be able to extend your existing topic hierarchy to accommodate those new messages.

- **Identifiers are useful in topic names:** Using a unique identifier for a device inside your topic hierarchy makes it easier to implement authorization through access control lists. Most brokers implement various authorization mechanisms, although those are not defined in the MQTT specification.

- **Be careful with the topic separator (forward slash):** In a topic name, do not use a forward slash as the leading or trailing character. Section 4.7.3 of the MQTT specification v5.0 states that "A leading or trailing '/' creates a distinct Topic Name or Topic Filter" and that "A Topic Name or Topic Filter consisting only of the '/' character is valid." This means that `sensors/humidity`, `/sensors/humidity`, `sensors/humidity/`, and `/sensors/humidity/` are all seen as distinct topics by the broker. Avoid the potential confusion and keep things simple. Moreover, using a leading slash introduces a useless zero-character level as the first level of the topic hierarchy. This is not LwM2M, dear readers!

Note   Many brokers expose server-specific information or control APIs under the `$SYS/` prefix, although the MQTT specification does not mandate this. Topics for which the name starts with $ will not match subscriptions starting with a wildcard character (# or +).

Quality of Service

Rather than the basic reliability mechanisms found in CoAP, MQTT offers three levels of delivery guarantees between a single sender and a single receiver. Those three levels of delivery guarantees are as follows:

- **At most once (QoS 0):** The message is delivered on a best-effort basis. The receiver will get the message once or not at all. Message loss can occur; the sender will not retry to send the message. The receiver will not send a response. This level is appropriate when data loss is acceptable, such as when reporting sensor readings in a non-mission-critical context.

- **At least once (QoS 1):** The message is delivered once at a minimum, but duplicates can occur. This level works well when you need guaranteed delivery but do not mind receiving multiple copies of the same message.

- **Exactly once (QoS 2):** The message will be delivered only once. However, the latency is not guaranteed; the message could be delayed for minutes or even hours due to congestion, networking issues, or outages on the client or broker side.

Publishers and subscribers specify their quality of service independently from each other. When the two do not match, the broker will downgrade the quality of service for the whole interaction. If a client publishes with QoS level 2 to a topic, subscribers who specified QoS level 1 in their subscription will receive the message at least once. In other words, it is possible subscribers get duplicates of the message even though it was published with the "exactly once" delivery guarantee.

Under QoS 1 and 2, if subscribing clients are offline for an extended time, queued messages will accumulate on the broker. In high-throughput environments, this means the memory and storage resources available to the broker could be overwhelmed. The MQTT specification does not specify how long queued messages should be kept – although MQTT v5 introduces a message expiry property. However, you can configure most brokers to enforce limits on the resources dedicated to queued messages. Eclipse Mosquitto, for example, can limit the number of queued messages or the number of bytes allocated to queued messages. Messages above those thresholds will be dropped silently. You can also impose a hard limit on Mosquitto's memory use, which will result in dropped messages and disconnected clients if exceeded.

Overall, higher quality of service increases latency and resource consumption, limiting the overall scalability of the infrastructure. Only select QoS 1 or 2 if you need their benefits.

Publish and Subscribe

When MQTT clients wish to publish a message, subscribe to a topic, or unsubscribe, they send specific packets.

To publish, clients need to send a PUBLISH packet. The most common properties for this packet are as follows:

- **packetId:** Unique identifier for a message. It is used in the context of delivery guarantees and will often be set for you by the MQTT client library or broker.

- **topicName:** Full name of the topic the message is published to.

- **qos:** A number between 0 and 2. It indicates the quality of service level for the message.

- **retainFlag:** Flag indicating whether the message will be saved as the last value for a specific topic. I will discuss retained messages later.

- **payload:** The actual content of the message. The MQTT specification states that the maximum message size is 268,435,456 bytes, which translates to 256 mebibytes (MiB) or 268 megabytes (MB). However, such large message sizes are not practical, and you should stick to slimmer payloads if you wish your solution to scale.

- **dupFlag:** Indicator for whether the message is a duplicate or not. This is only relevant to specific delivery guarantees. The MQTT client library or broker will handle it.

Under QoS level 1, the sender will store the message until it gets a PUBACK packet back acknowledging delivery. Correlation between the PUBLISH and PUBACK packets relies on the packetId property. However, PUBLISH packets will only be resent if the current TCP connection breaks and a new one is established. Clients can process the messages they receive under QoS level 1 immediately.

The flow for QoS level 2 is more complex. Once again, the packetId property of the original PUBLISH packet is used for correlation. Packet identifiers are unique between a specific client and a broker in the scope of any particular interaction. They can be reused once the interaction has concluded. The maximum value for the packetId property is 65535, and 0 is not allowed.

Publishing a message under QoS 2 involves the following four steps:

1. The sender transmits a PUBLISH packet with the qos property set to 2. It will store it until step 2 is completed.

2. The receiver gets the PUBLISH packet and processes it. It stores the packetId and replies with a PUBREC packet to acknowledge reception. The PUBLISH packet with be resent with the dupFlag set to true, but only if the current TCP connection breaks and a new one is established.

3. The sender discards its copy of the PUBLISH packet upon reception of the PUBREC. The sender stores the PUBREC. The sender transmits a PUBREL packet to authorize processing (release) of the packetId. The PUBREL will also be resent on session startup if a PUBCOMP hadn't been received before the connection was closed.

4. Upon reception of the PUBREL, the receiver discards the packetId and replies with a PUBCOMP (complete) packet. Any duplicate PUBLISH packet received until the receiver gets the PUBREL will not be processed. When the sender receives the PUBCOMP, it discards all the state information related to the interaction. At this point, the packetId can be reused for a new interaction.

To subscribe, clients need to send a SUBSCRIBE packet. Such packets enable you to specify a list of topic filters to subscribe to and, for each topic filter, a distinct level of quality of service. SUBSCRIBE packets also contain a packetId property, like PUBLISH packets.

The broker will process each requested subscription when it receives a SUBSCRIBE packet. Once done, it will send a SUBACK acknowledgment packet to the client. This message will contain the same packetId as in the SUBSCRIBE packet for correlation purposes. It will also feature a list of return codes for each of the subscriptions requested – in the same order. Table 4-1 lists the codes and their meanings.

Table 4-1. *MQTT subscription return codes*

Return Code	Meaning
0	Subscription successful; maximum QoS level is 0
1	Subscription successful; maximum QoS level is 1
2	Subscription successful; maximum QoS level is 2
128	Subscription failed

Subscription failures usually stem from malformed topic names or the lack of proper access rights.

Clients can unsubscribe from any topic at any time by sending an UNSUBSCRIBE packet to the broker. UNSUBSCRIBE packets contain a packetId and a list of topics to unsubscribe from. The broker will send back a UNSUBACK packet having the same packetId to acknowledge the unsubscribe request.

Connections and Sessions

Before sending or receiving messages, MQTT clients need to establish a connection with a broker. MQTT connections are kept open until the client sends a disconnection message unless network issues break the connection. Because the connection is kept open, MQTT clients will connect to brokers even if placed behind NAT infrastructure.

By default, brokers do not persist session information. This means subscribers need to reestablish their subscriptions every time they connect. For constrained devices, this is a waste of resources. Fortunately, the MQTT specification defines the concept of *persistent sessions*. When a client connects using a persistent session, the broker will keep track of all the subscriptions for that client. Moreover, it will store relevant messages when specific quality of service levels are involved:

- QoS 1 or 2 messages published while the client was offline

- QoS 1 or 2 messages published but not yet confirmed by the client

- QoS 2 messages published but not yet completely acknowledged by the client

Persistent sessions are useful but consume resources on the machine hosting the broker. You should use them only if you need them. Here is a list of valid reasons to do so:

- Your solution involves constrained clients that could extend their battery life or consume less bandwidth by relying on persistent sessions.

- Your use case requires subscribing clients not to miss a single message.

- Your use case involves "at least once" (QoS 1) or "exactly once" (QoS 2) message delivery.

Conversely, you should avoid persistent sessions for clients that just publish (no subscriptions) when the use case tolerates messages to be dropped.

Once a session is established, the client and broker will work behind the scenes to ensure it stays active through a *keep-alive* flow. Clients are supposed to send PINGREQ packets to the broker within a defined keep-alive interval; the broker must reply with a PINGRESP packet. Both packet types are devoid of a payload. Certain satellite and cellular links are known to break TCP/IP socket connections in ways not necessarily noticeable to the network stack. MQTT's keep-alive feature is useful in such environments. However, it is possible to deactivate it if desired. If you choose to leverage it, you should tweak the value of the keep-alive interval according to your devices and network constraints. Higher values will help you overcome fluctuations in signal strength or other causes of packet loss.

MQTT defines a Last Will and Testament (LWT) feature that enables you to notify subscribers that a specific publisher client has unexpectedly disconnected. This is done through a message. The client needs to provide the payload and topic for the message at connection time to do so. The broker will store the message and publish it to the topic specified once an unexpected disconnection occurs. The broker will discard the message if the client disconnects properly by sending a DISCONNECT packet. The following is a list of situations that will trigger the sending of a client's LWT:

- The broker detects an I/O error or network failure.

- The client fails to send a message within the defined keep-alive interval.

- The client omits to send a DISCONNECT packet before closing the network connection.

- The broker receives invalid packets from a client and forcefully closes the connection.

An MQTT client will send a packet of type CONNECT to a specific broker identified by its IP address or DNS hostname to establish a connection. CONNECT packets typically set the following properties in the variable header:

- **clientId:** Client identifier, which the broker can use to match a specific client to its state. It should be unique. In MQTT v5.0, it can be left empty, in which case the broker will assign one to the client and return it in the CONNACK packet. In MQTT v.3.1.1, it is possible to leave it empty if the connection was not persistent.

- **cleanSession:** Property indicating whether the broker should restore the session information for the provided clientId or not. If set to true, the session is nonpersistent, and any information for a previous persistent session will be purged. If set to false, the session is persistent; previous subscriptions will be restored, and queued messages will be delivered, acknowledged, or confirmed.

- **username/password:** Optional properties used for authentication. MQTT will transmit the values in plain text; consequently, you should at a minimum use TLS to ensure they are kept safe. MQTT version 5.0 introduced an enhanced authentication workflow supporting the implementation of challenge/response authentication.

- **lastWillTopic/lastWillMessage/lastWillQos:** Properties used to set the topic, payload, and QoS level for a publisher client's Last Will and Testament message.

Retained Messages

MQTT brokers can only distribute messages to subscribers if they are connected. By default, if a client subscribes to a topic, it will only get the next message to be published. For some use cases, this is undesirable. Retained messages allow new subscribers to get a copy of the last message published to a topic.

To create a retained message for a topic, a publisher needs to send a PUBLISH packet with the retainFlag property set to true. The broker will store the message and the corresponding QoS for the topic. Clients that subscribe to the topic after the retained message was sent will get a copy of the message as soon as they subscribe. This eliminates uncertainty about the last known good value for the topic. You must remember that if the publisher sends PUBLISH packets with retainFlag set to false after sending a retained message, the retained message for the topic will not change. The consequence is that newly subscribed clients will get a copy of the retained message, but not the others sent before establishing the subscription.

Retained messages, by the way, are not tied to persistent sessions. They will be delivered to subscribers independently of the session type chosen. It is possible to clear a retained message by sending a message with a zero-length payload with the retainFlag property set to true.

Protocol Stack

In the late 1990s, domain-specific proprietary field buses like CAN and Modbus were supplanted by TCP/IP-based alternatives such as Modbus TCP, Profinet, and OPC UA. Like those, MQTT supports TCP/IP. However, the specification does not mandate its use, rather stating that clients or servers must support "transport protocols that provide an ordered, lossless, stream of bytes from the Client to Server and Server to Client."[2] Figure 4-1 illustrates the protocol stack for MQTT.

---

[2]https://docs.oasis-open.org/mqtt/mqtt/v5.0/os/mqtt-v5.0-os.html#_Network_Connections

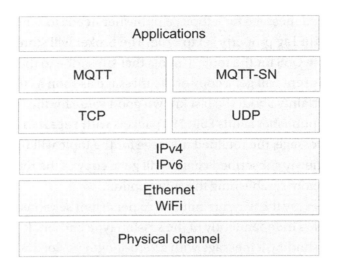

Figure 4-1. *MQTT and MQTT-SN protocol stack*

The MQTT specification mentions in a non-normative comment that TLS and WebSocket are also suitable transports. This is expected, as both protocols leverage TCP/IP. Nearly all brokers support TLS, and several implement support for WebSocket. This is the case of the two brokers hosted at the Eclipse Foundation: Eclipse Mosquitto and Eclipse Amlen.

Since it provides ordered and lossless delivery capabilities, TCP/IP is less suitable for battery-operated constrained devices operating over a wireless network than UDP. This is where MQTT-SN comes into play.

MQTT-SN

MQTT-SN can be considered a version of MQTT optimized for wireless networks and targeted at battery-operated devices. At the time of writing, MQTT-SN was undergoing standardization at OASIS Open. The starting point for this effort is version 1.2 of MQTT-SN, published in 2013 by IBM.

MQTT-SN supports UDP. It was originally developed for running on top of Zigbee networks. Such networks rely on IEEE 802.15.4 as the PHY and MAC layers protocol. A fundamental design objective was to reduce the size of messages with an eye on reducing power and bandwidth consumption.

At the architecture level, MQTT-SN defines three types of components: *clients*, *gateways*, and *forwarders*. MQTT-SN clients connect to an MQTT broker through an MQTT-SN gateway, which can be integrated into the MQTT broker or run as a stand-alone component. If the gateway is stand-alone, it communicates with the broker using MQTT. There are two types of gateways: *transparent* and *aggregating*. Transparent gateways create an MQTT connection to the broker for each connected client. Aggregating gateways use a single MQTT connection to the broker, which helps the broker scale when many MQTT-SN devices are involved. As for forwarders, they encapsulate the MQTT-SN frames received from clients on a wireless network and transmit them to a gateway. They are useful to bridge traffic between networks when it is impractical to have the clients and gateway on the same network.

This is a list of the main differences between MQTT and MQTT-SN:

- **CONNECT packet:** The CONNECT packet has been split into three. The two new ones enable the LWT topic and message transmission, respectively.

- **Topic IDs:** A 2-byte topic ID replaces the topic name in most messages. A registration procedure exists to enable clients to register their topic names, obtain an ID in return, and query for the ID of a topic.

- **Predefined topic IDs and short topic names:** Clients, gateways, and brokers can agree on the mapping between a topic ID and an actual MQTT topic. Consequently, they can use those topic IDs with a predefined meaning without registering them. As for short topic names, they are simply topic names that have a fixed length of two octets (16 bits). They can be used in place of topic IDs.

- **Discovery procedure:** Clients can discover the network address of a gateway or broker. Gateways periodically broadcast an ADVERTISE packet to make their existence known, and clients can send a SEARCHGW packet to obtain the ID of a gateway.

- **Persistent LWT:** The topic and message defined in the context of the Last Will and Testament feature are stored when a persistent session is used.

- **Offline keep-alive:** Sleeping clients are now supported. Messages destined to them are buffered on the gateway or broker and delivered when the client returns online.

- **QoS level -1:** This is a feature intended for minimal implementations. There are no connection, no registration, and no subscriptions. The client simply publishes its messages to a gateway for which it knows the address but does not check if the address is accurate or even if the gateway is alive. It is also possible to use a multicast address.

From a developer's perspective, MQTT-SN feels like plain MQTT. However, the protocol has seen much lower adoption than MQTT itself. Consequently, the software ecosystem is much more limited. The Eclipse Mosquitto team maintains the Really Small Message Broker (RSMB), a lightweight broker supporting MQTT and MQTT-SN. The Eclipse Paho project provides an MQTT-SN transparent gateway that can be used with Eclipse Mosquitto. Despite this, you definitively should have a look at MQTT-SN if you are deploying battery-operated constrained devices.

Note   Historically, RSMB was the inspiration for Mosquitto. Both code bases were contributed to the Eclipse Foundation by IBM and Roger Light, respectively.

New Features in MQTT v5

I mentioned earlier that the current version of MQTT, version 5.0, was introduced in 2019. Although there is a big jump in the version number, the new version is an evolution of MQTT 3.1.1 more than anything else. In this section, I will provide an overview of most of the new features that have been introduced in MQTT v5.0. Appendix C of the specification provides a comprehensive list. Some brokers had already implemented some of those features before OASIS Open formalized them in the specification.

I strongly recommend you use MQTT v5.0 wherever possible. Only utilize previous versions if you need to integrate legacy software or devices that absolutely cannot be upgraded to support version 5.

Note   You probably noticed there is no MQTT version 4. Why is that? The variable header for MQTT CONNECT packets contains a 1-byte unsigned value to represent the version of the protocol. MQTT v3.1 was assigned the value "3" and v.3.1.1 the value "4". MQTT v5 uses a value that matches its name. This may seem a boring bit of trivia, but it is more entertaining than explaining why there was never a Windows 9. The MQTT standardization committee decided that for future versions, the byte on the wire will always match the version of the protocol.

User-Defined Properties

Clients and brokers can now add an arbitrary number of custom headers to packets. Those custom headers are defined through a new key-value data type, where both key and value are represented as UTF-8 strings.

The MQTT specification does not assign a specific meaning to user-defined properties. It is up to the clients and brokers to decide what to do with them.

Reason Codes

In previous versions of MQTT, error handling was not very granular. For example, a client would know that its subscription request failed but would not be provided an exact cause for it. MQTT v5.0 changes that. In v5.0, many MQTT packets now include a predefined code that allows the client or broker to get context about error conditions. The following packets can carry reason codes:

- CONNACK

- PUBACK/PUBREC

- PUBREL/PUBCOMP

- SUBACK/UNSUBACK

- AUTH

- DISCONNECT

There are over 20 reason codes defined in the specification, with "Quota Exceeded" and "Protocol Error" as typical examples.

Availability of Optional Features

Some of the features defined in the MQTT specification are optional. Several brokers and *Software-as-a-Service* (SaaS) IoT offerings do not implement some of those optional features, including QoS 2, retained messages, and persistent sessions. For that reason, MTTQ v5.0 defines a way for partial implementations to signal to clients that they do not support specific features. The features concerned are Maximum QoS, Retain Available, Wildcard Subscription Available, Subscription Identifier Available, and Shared Subscription Available.

Message Expiry

This feature allows publishers to set an interval for message expiration. Brokers need to delete expired messages for which they did not start delivery.

Request/Response

This feature is a formalization of the request/response pattern in MQTT.[3] MQTT v5.0 defines properties that clients and brokers can leverage to implement request/response interactions: Response Topic, Correlation Data, Request Response Information, and Response Information. The specification describes how the properties can be used in a handful of non-normative sections.

Clean Session

The *Clean Session* flag has been split into two: a Clean Start flag indicating whether the session should reuse an existing session and a Session Expiry Interval indicating how long to retain the session after a disconnect. The Session Expiry Interval can be changed when sending a DISCONNECT packet. Setting Clean Start to 1 and Session Expiry Interval to 0 is equivalent in MQTT v3.1.1 of setting Clean Session to 1.

---

[3] https://docs.oasis-open.org/mqtt/mqtt/v5.0/os/mqtt-v5.0-os.html#_Toc3901252

The practical effect of this change is that in MQTT v5.0, sessions are nonpersistent by default since if the `Session Expiry Interval` is not specified in the `CONNECT` packet, then it will be set to zero, resulting in a nonpersistent session. Clients need to explicitly set the Session Expiry Interval to a value other than 0 if they wish to obtain a persistent session.

Server DISCONNECT

Brokers can now inform clients of an imminent disconnection by sending a `DISCONNECT` packet to them. Such packets will usually contain a reason code for the disconnection, making troubleshooting easier.

AUTH Packet

The new `AUTH` packet supports current challenge/response security flows, such as Kerberos or SCRAM. You can also leverage it in the context of token-oriented flows such as OAuth.

AUTH packets can be used to reauthenticate clients that are currently connected. Previously, the client would have had to close the connection to obtain the same result.

Passwords Without Usernames

AUTH packets are quite flexible, but it is still possible to implement token-oriented security flows without them. To make this a bit easier, it is now possible to specify a password in a `CONNECT` packet without providing a username.

Shared Subscriptions

Clients can now share the same subscription. Effectively, shared subscriptions represent a way to implement load balancing for MQTT subscriber clients.

To make this possible, MQTT v5.0 introduced the concept of *subscription groups*. Subscription groups are simply a group of clients sharing one or several subscriptions.

To use shared subscriptions, clients need to specify the topic this way:

```
$share/GROUPID/TOPIC
```

The three elements in the structure are as follows:

- $share, the predefined shared subscription identifier

- The group identifier, which is arbitrary and cannot contain wildcards

- The actual topic, which may include wildcards

Let me illustrate with a concrete example based on the Musée du Louvre once again. If you remember, I created an imaginary deployment of sensors in every room of the museum where the topic hierarchy was structured like this:

```
/Louvre/floor/room/sensor
```

Given this, a wildcard subscription for everything located on the second floor would be

```
Louvre/2/#
```

Suppose most of the rooms in the museum are located on this floor and that you wish to have several clients subscribing to the topic given the high number of messages published. If the name of the subscription group is LouvreSecondFloor, then the topic for the shared subscription would be

```
$share/LouvreSecondFloor/Louvre/2/#
```

The specification does not mandate a specific way for brokers to select which client in the subscription group will get a certain message. A round-robin approach is typical, but others are possible.

Flow Control

Brokers and their clients can specify the number of in-flight reliable messages (QoS > 0) they allow. This is a kind of rate-limiting mechanism. Whoever is the sender will pause message transmission to stay below the allowed threshold.

LWT Delay

MQTT v5.0 introduced the ability to specify a delay before the LWT message is sent in case of an unplanned disconnection. The LWT message is not sent if the connection is reestablished before the delay expires. This reduces the load on the MQTT infrastructure in environments where brief network interruptions are frequent.

Broker Reference

Brokers can specify an alternate broker when sending `CONNACK` and `DISCONNECT` packets. For example, this can be leveraged for provisioning purposes or to perform load balancing among a group of brokers. You can see the broker reference as a form of redirect.

Security

The core MQTT protocol contains few security features. This is not because the authors of the specification do not care about security but rather to give developers working on brokers and client libraries the opportunity to implement the security features that make sense for them. This also has the advantage of keeping the MQTT specification simple, since security topics are usually complex. The specification document for MQTT v.5.0 contains a non-normative section dedicated to security.

By default, MQTT does not encrypt communications. This means that in most cases, you should be using MQTT over TLS. Granted, TLS comes with an overhead due to the handshakes it requires and the larger size of encrypted packets on the wire. However, long-lived TCP connections mitigate the overhead of handshakes by making those less frequent. Moreover, *TLS session resumption* allows the reuse of a previously negotiated TLS session, reducing further the time and resources required when a client reconnects. Even with this, very constrained devices may have trouble coping with the CPU and bandwidth usage of TLS. In those extreme cases, you should, at a minimum, encrypt the payloads of `PUBLISH` messages and the passwords transmitted in `CONNECT` messages. Leveraging the `AUTH` message and challenge/response authentication is also an option.

If you decide to use TLS, make sure you always run the latest version supported by your broker and clients and validate the certificate chain. Certificates signed by an internal or development CA are fine for development and experimentation, but you should never deploy those in production environments. A trusted certificate authority should have signed certificates used in production. In addition, you should keep your eye open for any vulnerabilities found in the cipher suites you are using for TLS. In general, migrating to a more secure suite is well worth the effort.

Most brokers and clients integrate with IT infrastructure typically deployed to handle authentication and authorization, such as LDAP and OAuth authorization servers. The new `AUTH` packet added to MQTT v5.0 simplifies those types of integrations. You should always study the documentation for your broker and client libraries of choice to understand what authentication technologies they support and which authorization features they possess. Also, you should pay attention to the throttling features offered by your broker, which could help you mitigate the effects of attacks conducted by rogue clients to overload the broker. Limiting message size is also a good idea, if possible.

Broker: Eclipse Mosquitto

Enough theory; time to put our knowledge of MQTT to work by experimenting with Eclipse Mosquitto, a lightweight broker small enough to run on single-board computers. Mosquitto has been written in C and, as such, is highly portable. Mosquitto is available under the Eclipse Distribution License v1.0 and the Eclipse Public License v2.0.

The official web resources for Mosquitto are as follows:

- **Website:** https://mosquitto.org

- **Eclipse project page:** https://projects.eclipse.org/projects/iot.mosquitto

- **Code repository:** https://github.com/eclipse/mosquitto

At the time of writing, the most recent version of Mosquitto was 2.0.14, released on November 17, 2021.

Note   There is a second MQTT broker in the Eclipse IoT family: Eclipse Amlen. IBM contributed the code source of its IBM Watson IoT Platform Message Gateway to start the project. Amlen supports MQTT v3.1.1 and v5.0, in addition to JMS and custom protocols. Its footprint is larger than Mosquitto's, but it can be configured for clustering and high availability.

Installation

Mosquitto is available for Microsoft Windows, macOS, and most distributions of Linux. Compiling from source is also possible, of course.

You can download an installer from the Mosquitto website if you run Windows. 32-bit and 64-bit versions are available, and they will run on Windows Vista and higher.

If you are on macOS, Mosquitto can be installed from Homebrew. If it is already installed, all you need to do is to open a terminal window and issue the following command:

```
brew install mosquitto
```

There are various ways to install Mosquitto on Linux; some of them may not be available on your distro of choice. If you run Debian or one of its derivatives, such as Ubuntu, Raspbian, or the Raspberry Pi OS, Mosquitto is part of the official repositories. You can install it with this command:

```
 sudo apt install mosquitto
```

If you wish to get the latest versions more quickly, the Mosquitto team maintains a Debian repository that you can add to your sources.

Mosquitto is also available in the Fedora repository. If you use the Workstation or Server edition, this command will do the trick:

```
sudo dnf install mosquitto
```

Fedora IoT, on the other hand, does not rely on dnf. Packages are managed through rpm-ostree:

```
sudo rpm-ostree install mosquitto
```

If your distribution supports the snap packaging format, you can also install it that way:

```
snap install mosquitto
```

Configuration

Mosquitto works out of the box. The broker will start in local-only mode if no configuration file is provided. This means clients deployed on machines other than yours will not access it. The intent behind this behavior is to force you to build a proper configuration before using Mosquitto in production environments.

To start the broker once installed, access a command prompt, and issue this command:

```
mosquitto
```

If everything is set up properly, you should see output resembling this:

```
1640704397: mosquitto version 2.0.14 starting
1640704397: Using default config.
1640704397: Starting in local only mode. Connections will only be possible
from clients running on this machine.
1640704397: Create a configuration file which defines a listener to allow
remote access.
1640704397: For more details see https://mosquitto.org/documentation/
authentication-methods/
1640704397: Opening ipv4 listen socket on port 1883.
1640704397: Opening ipv6 listen socket on port 1883.
1640704397: mosquitto version 2.0.14 running
```

So far, so good. I will now perform a sanity check by subscribing to the system metrics topic exposed by Mosquitto. To do so, I will use the mosquitto_sub simple client developed by the Mosquitto team. The team also built mosquitto_pub, which you can use to publish to MQTT v3 and v5 brokers. Both utilities support TLS communications and leverage most MQTT features such as QoS and retained messages. On Debian and its derivatives, mosquitto_pub and mosquitto_sub are not part of the core mosquitto package but are found in the mosquitto_clients package instead.

Open a second command prompt on the machine where you installed mosquitto, and execute the following command:

```
mosquitto_sub -h localhost -p 1883 -t \$SYS/# -C 20 -v
```

The -C switch limits the number of messages received before the command exits, and -v prints the topic before the message payload. The output should look like this:

```
$SYS/broker/version mosquitto version 2.0.14
$SYS/broker/uptime 22 seconds
$SYS/broker/clients/total 0
$SYS/broker/clients/inactive 0
$SYS/broker/clients/disconnected 0
...
```

If you go back to the prompt where you launched Mosquitto, you will find lines that reflect the connection you just established in the log:

```
1640704494: New connection from 127.0.0.1:39616 on port 1883.
1640704494: New client connected from 127.0.0.1:39616 as auto-56DAB577-
D439-8D76-09BA-C3094A714D74 (p2, c1, k60).
```

The utility generates client IDs automatically. I could have specified one by using the -i switch.

Up to now, I have just used the default configuration for Mosquitto. To make the broker reachable on my local network, I need to create a configuration file and pass it to the Mosquitto process. The format for the configuration file is documented at length on the Mosquitto website. One thing to remember is that lines starting with "#" are treated as comments.

The following is a basic configuration that will make ports 1883 and 8883 reachable on all network interfaces. You are probably asking yourself where the files referenced by the cafile, keyfile, and certfile come from. I will explain how to create those in the next section.

```
# Standard MQTT traffic on port 1883, all interfaces
listener 1883
protocol mqtt

# MQTT over TLS traffic on port 8883, all interfaces
listener 8883
protocol mqtt
require_certificate true

cafile /etc/mosquitto/ca_certificates/mosquitto_ca.crt
keyfile /etc/mosquitto/certs/mosquitto_server_nopass.key
certfile /etc/mosquitto/certs/mosquitto_server.crt
```

To use this config file, copy the contents to a file (I chose /etc/mosquitto/mosquitto.conf) and launch Mosquitto this way:

```
mosquitto -c ./mosquitto.conf
```

If you wish to allow anonymous access to the broker, you can add this line at the top of the configuration file:

```
allow_anonymous true
```

Setting Up TLS

To use TLS, you will need a CA certificate plus a key and certificate for the broker. Self-signed certificates (not validated by a Certificate Authority) will not work. Moreover, you should make sure keys are not encrypted or password protected. Different brokers could have different requirements for the certificates. Always read the documentation attentively to generate certificates and keys suitable for your preferred broker.

The first step of the process is to generate a certificate and key for a certificate authority. This is the command you need to run:

```
openssl req -new -x509 -days 365 -extensions v3_ca -keyout mosquitto_ca.key -out mosquitto_ca.crt
```

You will need to answer several questions before the artifacts are generated. The following is an example:

```
Generating a RSA private key
...+++++
.........................................................+++++
writing new private key to 'mosquitto_ca.key'
Enter PEM pass phrase:
Verifying - Enter PEM pass phrase:
-----
You are about to be asked to enter information that will be incorporated
into your certificate request.
What you are about to enter is what is called a Distinguished Name or a DN.
There are quite a few fields but you can leave some blank
For some fields there will be a default value,
If you enter '.', the field will be left blank.
-----
Country Name (2 letter code) [AU]:CA
State or Province Name (full name) [Some-State]:Ontario
Locality Name (eg, city) []:Ottawa
Organization Name (eg, company) [Internet Widgits Pty Ltd]:Phantom Thieves
Organizational Unit Name (eg, section) []:P5 Royal
Common Name (e.g. server FQDN or YOUR name) []:blueberrycoder.com
Email Address []:
```

The second step is to generate a key for our broker. This is the command:

```
openssl genrsa -out mosquitto_server.key 2048
```

This is the output you will get:

```
Generating RSA private key, 2048 bit long modulus (2 primes)
....+++++
..........+++++
e is 65537 (0x010001)
```

The third step is to create a certificate signing request using the broker key. Issue this command:

```
openssl req -out mosquitto_server.csr -key mosquitto_server.key -new
```

Once again, you will need to answer questions about the country, state, locality, and common name (CN). Make sure you specify the hostname or domain name of the broker. In this case, I picked fedora, the hostname of one of my trusty Raspberry Pis.

Now, I can use the certificate authority's key and certificate to create and sign the broker's certificate.

```
openssl x509 -req -in mosquitto_server.csr -CA mosquitto_ca.crt -CAkey
mosquitto_ca.key -CAcreateserial -out mosquitto_server.crt -days 365
```

The output should look like this:

```
Signature ok
subject=C = CA, ST = Ontario, L = Ottawa, O = Phantom Thieves, OU = P5
Royal, CN = fedora
Getting CA Private Key
Enter pass phrase for mosquitto_ca.key:
```

Whew! At this point, we have everything the broker needs. You could create the files on a development machine and use scp to transfer them to the machine hosting the broker.

Each of the clients will also need a certificate. The process to create those is similar to the one used for the broker. The instructions I provide create certificates suitable for mosquitto_sub and mosquitto_pub.

First, let's create a client key:

```
openssl genrsa -out mqtt_client.key 2048
```

You should see something like this:

```
Generating RSA private key, 2048 bit long modulus (2 primes)
..........................................................+++++
......................................+++++
e is 65537 (0x010001)
```

Let's now create a signing request.

```
openssl req -out mqtt_client.csr -key mqtt_client.key -new
```

Once again, you will need to answer a slew of questions. In this case, I created a certificate for my beloved Ashitaka workstation. Here is how things should look:

```
You are about to be asked to enter information that will be incorporated
into your certificate request.
What you are about to enter is what is called a Distinguished Name or a DN.
There are quite a few fields but you can leave some blank
For some fields there will be a default value,
If you enter '.', the field will be left blank.
-----
Country Name (2 letter code) [AU]:CA
State or Province Name (full name) [Some-State]:Ontario
Locality Name (eg, city) []:Ottawa
Organization Name (eg, company) [Internet Widgits Pty Ltd]:Phantom Thieves
Organizational Unit Name (eg, section) []:P5 Royal
Common Name (e.g. server FQDN or YOUR name) []:Ashitaka
Email Address []:

Please enter the following 'extra' attributes
to be sent with your certificate request
A challenge password []:
An optional company name []:
```

I sign the signing request with the CA's key and certificate.

```
openssl x509 -req -in mqtt_client.csr -CA mosquitto_ca.crt -CAkey
mosquitto_ca.key -CAcreateserial
 -out mqtt_client.crt -days 365
```

The output should resemble this:

```
Signature ok
subject=C = CA, ST = Ontario, L = Ottawa, O = Phantom Thieves, OU = P5
Royal, CN = Ashitaka
Getting CA Private Key
Enter pass phrase for mosquitto_ca.key:
```

You can use the client certificate to publish or subscribe on the broker at this point. It is essential to use the hostname for the broker and not its IP address; you may have to tweak your internal DNS server or add entries to /etc/hosts, among other things.

For example, you could subscribe like this:

```
mosquitto_sub -h fedora -p 8883 -t wisdom/# -v --cafile mosquitto_ca.crt --
cert mqtt_client.crt -i Ashitaka
```

To publish some wisdom, you could use this command:

```
mosquitto_pub -h fedora -p 8883 -t wisdom/Rome -m 'Carpe Diem' --cafile
mosquitto_ca.crt --cert mqtt_client.crt
```

Sandbox Servers

There is a publicly available instance of Mosquitto at test.mosquitto.org on ports 1883 and 8883. It supports plain MQTT, MQTT over TLS, MQTT over TLS (with client certificate), MQTT over WebSockets, and MQTT over WebSockets with TLS.

The Eclipse Paho team also maintains a Mosquitto sandbox. It can be found at mqtt.eclipseprojects.io on ports 1883 and 8883. It supports plain MQTT and MQTT over TLS (with client certificate).

Both servers always run the latest version. Please use them responsibly and do not count on them for production applications. There are no access controls configured on those servers, so any data you publish will be accessible to all.

Building Clients with Eclipse Paho

`mosquitto_pub` and `mosquitto_sub` are useful troubleshooting tools built on libmosquitto, a C library you can use to integrate MQTT support into your programs.

An alternative to libmosquitto is the Eclipse Paho collection of MQTT client libraries. Paho offers MQTT client libraries in the following languages: C, C# (.Net), C++, GoLang, Embedded C/C++ (for constrained devices), Java, JavaScript, Python, and Rust. There is also an Android service available. Each library is developed independently and supports MQTT 3.1.1 at a minimum. Paho offers many advanced features, such as automatic reconnect, offline buffering, high availability, and a choice of blocking and nonblocking APIs. Not all libraries support all possible features, but the Java, C, and C++ versions do.

Like Mosquitto, Paho is published under the Eclipse Distribution License v1.0 and the Eclipse Public License v2.0.

I will now show you how to leverage the Java and Python versions of Paho to build simple MQTT programs.

Java

The Java version of Paho relies on Apache Maven as its build system, and the binaries are published to Maven Central. Listing 4-1 shows you the dependency declaration you should add to your `pom.xml`.

Listing 4-1. Maven dependency declaration for Paho Java

```
<dependencies>
    <dependency>
        <groupId>org.eclipse.paho</groupId>
        <artifactId>org.eclipse.paho.mqttv5.client</artifactId>
        <version>1.2.5</version>
    </dependency>
</dependencies>
```

Please note that the artifactId for the MQTT v3 version of the library is `org.eclipse.paho.client.mqttv3`.

Listing 4-2 is some sample code that illustrates how to publish a message using the blocking API. It is inspired by the sample found in the README.md for the library's repository. I used the IP address for my broker here, but you could substitute mqtt. eclipseprojects.io or test.mosquitto.org.

Listing 4-2. Simple Paho Java publishing client

```java
String topic       = "wisdom/Rome";
String content     = "Omnium Rerum Principia Parva Sunt";
int qos            = 2;
String broker      = "tcp://192.168.2.16:1883";
String clientId    = "JavaSample";
MemoryPersistence persistence = new MemoryPersistence();

try {
    MqttClient sampleClient = new MqttClient(broker, clientId,
    persistence);
    MqttConnectionOptions connOpts = new MqttConnectionOptions();
    connOpts.setCleanStart(true);
    System.out.println("Connecting to broker: "+broker);
    sampleClient.connect(connOpts);
    System.out.println("Connected");
    System.out.println("Publishing message: "+content);
    MqttMessage message = new MqttMessage(content.getBytes());
    message.setQos(qos);
    sampleClient.publish(topic, message);
    System.out.println("Message published");
    sampleClient.disconnect();
    System.out.println("Disconnected");
    System.exit(0);
} catch(MqttException me) {
    System.out.println("reason "+me.getReasonCode());
    System.out.println("msg "+me.getMessage());
    System.out.println("loc "+me.getLocalizedMessage());
    System.out.println("cause "+me.getCause());
    System.out.println("excep "+me);
}
```

The flow of the code is straightforward. Here is a list of the main steps in the preceding sample:

1. Create an instance of the `MqttClient` class. The constructor requires the broker's URI, a client ID, and an instance of a persistence class. The sample uses memory persistence, but file persistence is also available.

2. Create an instance of `MqttConnectionOptions` and set the options we need. In this case, we opt for a clean session (`setCleanStart(true)`).

3. Connect to the broker by calling the `connect` method of MqttClient, passing our `MqttConnectionOptions` as a parameter.

4. Create a message. The constructor for the `MqttMessage` class accepts the payload as a byte array only. This is why I call `getBytes()` on the `String` holding the payload.

5. Call the `publish` and `disconnect` methods on our instance of MqttClient.

If you need a more comprehensive sample, you could study the code of the Paho mqtt-client CLI application available in the GitHub repository for Paho's Java library.

Python

The Python version of Paho is fairly comprehensive. It offers blocking and nonblocking APIs and supports MQTT v3 and v5. However, it does not implement message persistence or high availability at the time of writing. There are also some limitations involving QoS 1 and QoS 2 messages. The library is available in the *Python Package Index* (PyPI), and you can install it using the `pip` dependency manager. Python 2.7.9+ or 3.6+ is required.

To install the library, execute this command:

```
pip install paho-mqtt
```

Let's now look at Listing 4-3: a little program that subscribes to a specific topic and prints payloads as messages come in.

Listing 4-3. Simple Paho Python publishing client

```python
import paho.mqtt.client as mqtt

# The callback for when the client receives a CONNACK response from
the server.
def on_connect(client, userdata, flags, rc, properties):
    print("Connected with result code "+str(rc))

    # Subscribing in on_connect() means that if we lose the connection and
    # reconnect then subscriptions will be renewed.
    client.subscribe("wisdom/#")

# The callback for when a PUBLISH message is received from the server.
def on_message(client, userdata, msg):
    print(msg.topic+" "+str(msg.payload))

client = mqtt.Client("PythonSample", protocol=mqtt.MQTTv5)
client.on_connect = on_connect
client.on_message = on_message

# Use the IP or hostname of your broker.
# You can also use the sandbox at mqtt.eclipseprojects.io
client.connect("192.168.2.16", 1883, 60)

# Blocking call that processes network traffic, dispatches callbacks and
# handles reconnecting.
# Other loop*() functions are available that give a threaded
interface and a
# manual interface.
client.loop_forever()
```

Once again, this sample is inspired by code available in the README file of the library's repository. I modified it to use MQTT v5. Let's now have a close look at the code's flow. The main steps in the preceding code are as follows:

1. Declare a callback named on_connect that will be called once the client connects to the broker. This callback contains the logic to subscribe to a topic (wisdom/Rome).

2. Declare a callback named on_message that will be called once a message is received. I just print the message's topic and payload.

3. Create an instance of the Client object. The two parameters in the constructor are a client ID and the version of MQTT I wish to use, v5 in this case. I also assign the two callbacks declared before.

4. Call the connect method on our Client instance. The last line ensures the program loops until interrupted (Ctrl+C).

If you paste the code in a .py file and execute it, you will see this if the connection is successful:

```
Connected with result code Success
```

Let's say I publish a message on the wisdom/Rome topic with mosquitto_pub:

```
mosquitto_pub -h 192.168.2.16 -p 1883 -t wisdom/Rome -m 'Timendi Causa Est
Nescire'
```

Then, the output will be

```
wisdom/Rome b'Timendi Causa Est Nescire'
```

MQTT and Constrained Devices

Constrained devices powerful enough to run a full TCP/IP stack can interact with MQTT brokers. Several open source and commercial libraries are available that target such devices, and some RTOSes ship with support for MQTT. If maximizing battery life is critical, going with MQTT-SN could be a better choice.

I will now explain how you can leverage the Eclipse Paho embedded clients and give you pointers to get started on the Zephyr RTOS.

Eclipse Paho Embedded Clients

The Eclipse Paho project maintains separate embedded libraries for MQTT and MQTT-SN. Both are built using a layered approach; they do not rely on particular dependencies for networking, threading, or memory management. Moreover, they are written in ANSI standard C to make the code portable.

The Paho Embedded MQTT client is made of three distinct components:

- **MQTTPacket:** A tiny library focused exclusively on the serialization and deserialization of MQTT packets.

- **MQTTClient:** A C++ library based on MQTTPacket that was initially written for the Arm mbed RTOS. It has since been ported to other platforms, such as Linux and Arduino. It avoids dynamic allocations and eschews the use of the STL.

- **MQTTClient-c:** A direct translation of MQTTClient written in C. It works on the FreeRTOS kernel and could be used in similar environments.

MQTTPacket supports all current versions of MQTT, and the other two support MQTT v3.1 and v3.1.1 only. To keep code size lower, they also removed some features compared to the full C and C++ implementations: message persistence, automatic reconnect, offline buffering, and WebSocket support. Moreover, MQTTClient and MQTTClient-C offer a blocking API only. The GitHub repository contains samples for the three components on various operating systems.

The Paho Embedded MQTT-SN client uses the same approach as its MQTT brethren. It supports MQTT-SN v1.2. The client contains two main components:

- **MQTTSNPacket:** A library to handle the serialization and deserialization of MQTT-SN packets. It also provides many helper functions.

- **MQTTSNClient:** A C++ library based on MQTTSNPacket. It is a work in progress.

The repository for the client also contains the code for the Paho MQTT-SN transparent gateway I already mentioned.

Zephyr RTOS

The Zephyr RTOS from the Linux Foundation ships with a built-in MQTT library; the implementation only supports MQTT v3.1.0 and v3.1.1. The library is built on the top of BSD sockets and supports TLS.

The Zephyr repository on GitHub contains a sample publishing client.

CHAPTER 5

Sparkplug

À certains moments de notre vie, notre propre lumière s'éteint et se rallume par l'étincelle d'une autre personne.

At certain times in our life, our own light is extinguished and rekindled by the spark of another person.

—Albert Schweitzer

Since 1999, MQTT has known tremendous growth and adoption. The inception of the Mosquitto and Paho projects at the Eclipse Foundation in 2012–2013 created an open, vendor-neutral community around the protocol. The most successful ecosystems are those that start and stay open.

As I have explained earlier, there are many things in MQTT that you will need to decide arbitrarily. The MQTT specification is silent on how you should structure your topic hierarchy and does not say anything about the format and encoding of the payloads. Do you want to use plain text? Fine! Base64 encoded images? Sure thing! JSON, YAML, or XML? Of course! And at this point in the discussion, you probably see the problem. Devices and applications relying on MQTT have no idea what topic to publish or subscribe to. As for the payloads, they can be and will be anything and everything. Out of the box, MQTT infrastructure is not interoperable.

Of course, you can design devices and applications to get the names of the topics through a configuration mechanism. You can even implement clever ways to massage payloads or implement custom logic to translate to and from a shared data model. But doing those things requires time. More importantly, the task will grow more complex as you add devices and applications. Fortunately, there is a better way: Eclipse Sparkplug.

Sparkplug is an open source specification hosted at the Eclipse Foundation *"[...] that provides MQTT clients the framework to seamlessly integrate data from their*

© Frédéric Desbiens 2023
F. Desbiens, *Building Enterprise IoT Solutions with Eclipse IoT Technologies*,
https://doi.org/10.1007/978-1-4842-8882-5_5

applications, sensors, devices, and gateways within the MQTT Infrastructure."[1] It is developed in the open, on GitHub. The evolution of Sparkplug is governed by the Eclipse Foundation Specification Process (EFSP). Describing at length the EFSP process is outside the scope of this book, but the main thing you should remember about it is that the documents produced are open and the software open source. Anyone can copy and distribute the specification documents in any medium for any purpose and without fee or royalty.

Note   Although you may not have heard about it before, the EFSP has been in use since 2018. The Jakarta EE specifications, which succeeded Java Enterprise edition, are managed under the EFSP.

Sparkplug is not some fangled newcomer in the market. The first version of the specification goes back to 2016 when Arlen Nipper and his team at Cirrus Link Solutions made it available. It has been widely adopted since, and many commercial and open source implementations are available. Eclipse Tahu is the Eclipse Foundation's open source implementation, but the Linux Foundation's Fledge project also has a Sparkplug connector available.

Sparkplug and Interoperability

The lack of interoperability inherent to MQTT is the problem Sparkplug is trying to solve. How does it tackle this issue? Let's see what the authors of the Sparkplug specification have to say:

> *The intent and purpose of the Sparkplug Specification is to define an MQTT Topic Namespace, payload, and session state management that can be applied generically to the overall IIoT market sector, but specifically meets the requirements of real-time SCADA/Control HMI solutions. Meeting the operational requirements for these systems will enable MQTT based infrastructures to provide more valuable real-time information to Line of Business and MES solution requirements as well.*[2]

---

[1] https://sparkplug.eclipse.org/about/faq/

[2] https://github.com/eclipse/sparkplug/blob/master/specification/src/main/asciidoc/chapters/Sparkplug_1_Introduction.adoc

The logic is this: if MQTT devices and applications share a topic namespace, state management, and payload, they will integrate simply and efficiently. Here is a high-level overview of those three concepts:

- **Topic namespace:** Sparkplug defines an extensible topic namespace optimized for the SCADA and Industrial IoT (IIoT) sector. This namespace allows automatic discovery and bidirectional communications between MQTT clients.

- **State management:** Sparkplug fully takes advantage of the "Continuous Session Awareness" capabilities of MQTT to reduce latency in scenarios where unreliable networks offering limited bandwidth are involved, making it a good choice for real-time SCADA and IIoT solutions. This is especially important because Sparkplug relies on a *report by exception* approach where there are no periodic updates provided to subscribers. Data updates are sent whenever they occur only, and this can be done reliably only because of Sparkplug's state management features.

- **Payload:** Sparkplug defines a binary message encoding that works well for legacy register-based process variables, such as Modbus register values. The payload is focused on process variable change events called *metrics*. The payload format supports complex data types, datasets, metrics, metric metadata, and metric aliases. The payload format is versioned and is referenced in the topic structure to ensure clients can easily infer the encoding used for a specific message.

Before I explain in detail Sparkplug's topic structure, state management, and payload, let's have a look at the architecture, requirements, and basic principles of the protocol.

Architecture

In addition to devices and sensors, the reference architecture for Sparkplug defines three types of components: MQTT Server, MQTT Edge Node, and Host Application. Figure 5-1 illustrates how these components are organized.

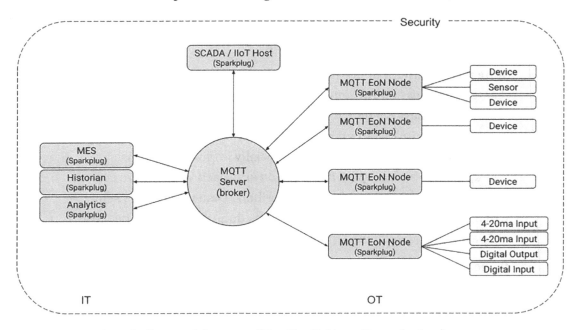

Figure 5-1. *Sparkplug architecture (Credit: Eclipse Foundation)*

The MQTT Server (broker) is the central component of the architecture. It needs to support a specific subset of MQTT features, which I will discuss in the following section. Sparkplug supports the use of multiple brokers to support high availability and redundancy requirements, which makes the infrastructure more scalable.

MQTT Edge Nodes are MQTT client applications that play a gateway role in the architecture. They implement local interfaces to legacy devices such as programmable logic controllers (PLCs), remote terminal units (RTUs), flow computers, and sensors. They often interface with local discrete I/O and logical internal process variables (PVs). Edge nodes enable non-Sparkplug components to participate in the Sparkplug topic namespace and use the Sparkplug payload format.

Sparkplug Host Applications are MQTT clients that consume the real-time Sparkplug messages. Host Applications must advertise that they are online or offline, as specified by the protocol. In typical SCADA or IIoT implementations, a single Primary Host

Application is responsible for monitoring and controlling a given edge node. The edge node indicates its Primary Host Application, and specific Sparkplug STATE messages are used to inform edge nodes whether the Primary Host is connected and subscribed to the MQTT Server. However, Sparkplug supports the notion of additional Host Applications in a monitoring or standby mode – whether their scope is the full set of edge nodes or a specific subset. Edge nodes can be configured to have distinct behaviors when their Primary Host Application is online or offline.

Sparkplug MQTT Requirements

For the time being, Sparkplug is explicitly tied to MQTT as its transport protocol. It could theoretically be ported to other protocols but relies upon specific behaviors of MQTT to implement some of its features.

Sparkplug supports both MQTT v3.1.1 and v5.0. It will work on any combination of broker and client library that fully implement the following features:

- Publish and subscribe with QoS levels 0 and 1

- Retained messages

- Last Will and Testament

Moreover, in a non-normative section on security, the specification recommends relying on Access Control Lists (ACLs) to restrict the topics a client can interact with and run MQTT over TLS. These recommendations can influence the MQTT broker or SaaS solution you select for your implementation.

Eclipse Mosquitto and Eclipse Amlen both fulfill the requirements listed previously and can be used to implement Sparkplug-based solutions.

Basic Principles

The Sparkplug specification highlights a few important design principles. Those principles are as follows:

- Publish/subscribe approach

- Report by exception

- Continuous session awareness

- Birth and death certificates

- Support for persistent and nonpersistent connections

The main interest of a publish/subscribe approach, inherited from MQTT, is the complete decoupling of devices from the applications consuming the data. In contrast to previous approaches to integration, which were point to point, the MQTT server (broker) in Sparkplug plays the role of a hub.

A key concern for constrained devices is reducing power and bandwidth consumption as much as possible. Report by exception is an elegant way to achieve that. In most cases, edge nodes only need to send messages when the values they report about change. This is possible because Sparkplug mandates stateful MQTT sessions. That said, developers are free to use periodic reporting when reporting by exception does not make sense.

A fundamental rule of Sparkplug is that edge nodes or Host Applications must advertise that they are now online or about to go offline. This is achieved using birth and death certificates. Death certificates are always registered in the CONNECT packet as the will message (LWT) of the node or application. As for birth certificates, the specification states they must be the first message published by any node or application.

Finally, Sparkplug lets developers decide for themselves if the edge nodes should be connected permanently to the MQTT infrastructure or if they should just connect as needed. In the case of periodic connections, all the nodes need to do is publish their own death certificate and then send a proper DISCONNECT packet to close the connection gracefully. This ensures that a death certificate will not be sent later by the MQTT Server. Host Application will then treat the metrics received from the node as "Last Known Good" values until it returns online and refreshes them. Regardless of their choice to be continuously connected or periodically connected, edge nodes must never use MQTT persistent sessions. In other words, edge nodes must always set the cleanSession (MQTT v3) or cleanStart (MQTT v5) flag to true.

Topic Namespace

The goal of the Sparkplug topic namespace is to organize the data received from edge nodes and devices logically yet concisely. The structure of the topic hierarchy looks like this:

```
namespace/group_id/message_type/edge_node_id/[device_id]
```

I will now explain the meaning of each of the elements in the hierarchy.

Namespace

The namespace element is a constant expressing the version of Sparkplug used. It signals to message consumers both the structure of the other levels in the topic hierarchy and the encoding of payload data. The namespace element at the time of writing looked like this:

```
spBv1.0
```

The structure employed is

```
sp[Payload encoding scheme][Payload encoding scheme version]
```

Consequently, the rest of this chapter describes version v1.0 of payload encoding scheme "B", also known as *Sparkplug B*.

Note   The very first version of the Sparkplug specification defined payload encoding scheme "A". However, implementers found some issues with it, and it was quickly replaced by payload encoding scheme "B". Payload encoding scheme "A" is discontinued and not documented in the specification anymore. Please do not use it.

Group ID

The group ID element of the topic namespace is used to group the edge nodes logically. It must be a valid UTF-8 MQTT string; as such, it cannot contain the topic filter wildcard characters ("+" and "#") nor the topic level separator "/".

The specification recommends using descriptive group ID names but keeping them as short as possible.

Message Type

The Sparkplug specification defines several message types. This mimics the way MQTT packets are defined. Due to its structure, the topic namespace segregates messages according to their type. I will discuss message types at length in the next section.

edge_node_id

This element is an ID for the edge node within the infrastructure. It must be a valid UTF-8 MQTT string. The specification states that the combination of the group ID and the edge node ID must be unique. It also recommends keeping edge node IDs as short as possible since they travel with every message published.

device_id

The device ID identifies a device attached to the edge node physically or logically. This is an optional element in the namespace since some messages target the edge node and not a specific device. The device ID must be unique in the scope of the edge node if it is used. It must be a valid UTF-8 MQTT string.

Devices with built-in support for Sparkplug could decide to present themselves as an edge node rather than a device. In that case, they would not produce or process device-related messages and would not use device IDs.

Message Types

Sparkplug defines and documents several types of messages. Using those, Host Applications

- Can discover the metadata and monitor the state of all the edge nodes and devices connected to the MQTT infrastructure

- Can discover all the metrics collected by the devices or edge nodes, including diagnostics, properties, metadata, and current state values

- May issue command messages to any edge node or Device metric if the implemented security model allows it

This is the full list of Sparkplug message types, along with a brief explanation of their contents:

- **NBIRTH:** Birth certificate for Sparkplug Edge Nodes. The message lists everything the edge node itself will report (node metrics) but not the device metrics. At a minimum, it will contain the metric name, data type, and current value for every metric.

- **NDEATH:** Death certificate for Sparkplug Edge Nodes. This message is set as the will message for the client upon connection. The payload contains a correlation identifier enabling the Host Application to match the birth and death certificates. Upon reception, the Host Application will mark the metric values for the node as stale.

- **NDATA:** Edge node data message. The payload of these messages contains the edge node metric values that need to be reported to subscribing clients because they have changed. There can be one or several metrics in the payload.

- **NCMD:** Edge node command message. Such messages are used to write a new value for a metric on an edge node.

- **DBIRTH:** Birth certificate for Devices. Host Applications will create or update the metric structure for the device only when they receive the relevant DBIRTH. In addition to value metrics, the DBIRTH message can contain definitions for device control metrics and device properties. Device control metrics are used for tasks such as rebooting or setting a scan rate. Device properties can document the manufacturer, model, and firmware version of a device, among other things.

- **DDEATH:** Death certificate for Devices. The edge node sends this message to signal the unavailability of a device to the Host Application. Upon reception, the Host Application will mark the metric values for the device as stale.

- **DDATA:** Device data message. The payload of these messages contains the device metric values that need to be reported to subscribing clients because they have changed. There can be one or several metrics in the payload.

- **DCMD:** Device command message. Such messages are used to write a new value for a metric on a device.

- **STATE:** Sparkplug Host Application state message. This message type is used for both the birth and death certificates of the Host Application. The birth certificate will be the first message sent by the Host Application when it comes online. It will be a retained message delivered over QoS level 1 with the UTF-8 string "ONLINE" as the payload. As for the death certificate, it simply substitutes the string "OFFLINE" as the payload and is registered as the MQTT Will Message for the Host Application in the MQTT CONNECT packet.

  STATE messages are published in a topic hierarchy of their own. This is the structure used: /STATE/sparkplug_host_application_id. The application ID is arbitrary but must be unique for every Sparkplug Host Application in the infrastructure.

Payload Definition

The payload definition for Sparkplug enables application builders and device makers to transmit values and their metadata in a very flexible way.

At a technical level, the authors of the Sparkplug specification decided on Google Protocol Buffers as their solution for structuring Sparkplug payloads and handling their serialization and deserialization. Protocol Buffers is a language-neutral, platform-neutral, extensible mechanism for serializing structured data. Protocol buffer data is structured using a domain-specific language and stored in .proto files. Google provides a compiler to generate data access classes from those files.

The Sparkplug specification contains an integral copy of the Sparkplug payload in .proto format. The Eclipse Tahu project also has a copy in its repository.

Note   Any timestamp property you will see used in Sparkplug is an unsigned 64-bit integer representing the number of milliseconds since the UNIX epoch (January 1, 1970, UTC).

The top-level structure of a payload expressed in JSON looks like this:

```
{
  "timestamp": 1641948773752,
  "metrics": [],
  "seq": 1,
  "uuid": "base64png",
  "body": "an array of bytes"
}
```

The uuid, which is optional, represents a schema or arbitrary information, enabling message recipients to parse the body property. The body is an array of bytes that can accept any binary encoded data.

Metrics

Sparkplug metrics are simply key/value/data type values. You can add optional metadata and key/value properties to them. Here is an example:

```
{
  "name": "exterior_temperature",
  "alias": 130870,
  "timestamp": 1641950175801,
  "datatype": "Int8",
  "is_historical": false,
  "is_transient": false,
  "is_null": false,
  "metadata": {},
  "properties": {},
  "value": 23
}
```

In this case, I left metadata and properties empty, so you could visualize where they go. The metadata property can describe binary data transferred through Sparkplug messages, especially files.

Sparkplug payloads can also contain DataSets and Templates. DataSets are used to encode data matrices. To define one, you need to provide the number of columns, their names, and their types in addition to the data itself. As for templates, they allow you to define your custom data types. This makes the Sparkplug payload encoding flexible and extensible.

Session Management

Session management is one of the core features of Sparkplug. The specification's authors explain why this feature is so important:

> *Due to the nature of real time SCADA solutions, it is very important for the Primary Host Application and all connected Eclipse Sparkplug Edge Nodes to have the MQTT Session STATE information for each other. In order to accomplish this the Sparkplug Topic Namespace definitions for Birth/ Death Certificates along with the defined payloads provide both state and context between the Primary Host Application and the associated Edge Nodes.*[3]

That said, the main difference between a Primary Host Application and any other Sparkplug Host Application is that the edge nodes will check that the Primary Host Application is online before starting to publish data. Note the Primary Host is identified by each edge node. For example, within a given Sparkplug infrastructure, one edge node's Primary Host may be different from another's.

Example

Suppose I wish to migrate the *Musée du Louvre* to Sparkplug. How would things look? As a reminder, I was reporting temperature and humidity values from rooms located on various floors. The topic structure for a sample sensor would be

`/Louvre/2/201/humidity`

Here, the two biggest factors are the number and location of our edge nodes. Do we have one node per room? One per floor? Given that we have only two sensors per room, I assume we have one edge node per floor. Given this, the Edge Node ID can represent the floor, and the device ID can convey the room.

---

[3] https://github.com/eclipse-sparkplug/sparkplug/blob/master/specification/src/ main/asciidoc/chapters/Sparkplug_5_Operational_Behavior.adoc

What about the group ID, then? In this case, I am deploying at the *Louvre,* but the *Musée des Beaux-Arts du Canada* (MBAC) is next. So I will use Louvre as the group ID.

All right. With that decided, let's see the topics for our Host Application, Edge Nodes, and Devices.

Host Application

Suppose the Host Application here is called GLaDOS. In this case, we could use this for the application ID. The topic will be

STATE/GLaDOS

Edge Nodes

I decided to have one edge node per floor and use the node ID to represent the floor number. The topics for my edge nodes will then follow this pattern:

spBv1.0/Louvre/message_type/floor

Or more concretely for the first floor's edge node:

spBv1.0/Louvre/NBIRTH/1
spBv1.0/Louvre/NDEATH/1
spBv1.0/Louvre/NDATA/1
spBv1.0/Louvre/NCMD/1

Devices

For the sake of this example, I will suppose that my temperature and humidity sensors are different devices connected to the edge node for their floor. To keep things simple, I will assume each of them exposes a single process variable. Real-world devices usually expose hundreds or even thousands of variables. I want to convey the room number through the device name. The topic pattern would be this:

spBv1.0/Louvre/message_type/floor/device

So for room 101, I would have this:

spBv1.0/Louvre/DBIRTH/1/temp-101

```
spBv1.0/Louvre/DDEATH/1/temp-101
spBv1.0/Louvre/NDATA/1/temp-101
spBv1.0/Louvre/NCMD/1/temp-101
spBv1.0/Louvre/DBIRTH/1/hum-101
spBv1.0/Louvre/DDEATH/1/hum-101
spBv1.0/Louvre/NDATA/1/hum-101
spBv1.0/Louvre/NCMD/1/hum-101
```

Now, the only thing we need to do is send some data over the wire. The DBIRTH payload could look like this:

```
{
    "timestamp": 1641940113424,
    "metrics": [ {
        "name": "temp-value",
        "alias": 1,
        "timestamp": 1641940113424,
        "dataType": "Int8",
        "value": -24,
        "transient": false,
        "null": false,
        "historical": false
    }],
    "seq" : 1,
    "metricCount" : 1
}
```

Here is a Sparkplug payload for the DDATA message of our temperature sensor:

```
{
    "timestamp": 1641940114576,
    "metrics": [{
    "name": "",
    "alias": 1,
    "timestamp": 1641940024394,
    "dataType": "Int8",
    "value": -25
```

```
  }],
  "seq": 1
}
```

Leveraging Eclipse Tahu

Eclipse Tahu is an open source implementation of Sparkplug developed at the Eclipse Foundation. It comprises five distinct client libraries in C, C#, Java, JavaScript, and Python, each with different levels of maturity and adoption. At the time of writing, the Java and C versions were the best maintained.

Note   The EFSP requires all Eclipse specifications to be accompanied by a compatible open source implementation, that is, an implementation that passes the *Technology Compatibility Kit* (TCK) for that specification. A TCK is software and documented requirements that support the testing of implementations to ensure that they are compatible with the specification. TCKs are built by the relevant Eclipse specification project team.

At the time of writing, the best way to get started with Tahu is to clone the repository and compile the version you are interested in yourself. The Java version uses Apache Maven as its dependency management system. To compile, go to the folder where you clone the repository and execute

```
mvn clean install
```

This will build the Java version as well as the Java sample applications. Once this is done, you can add this dependency in your own pom.xml:

```
<dependencies>
    <dependency>
        <groupId>org.eclipse.tahu</groupId>
        <artifactId>tahu</artifactId>
        <version>0.5.12</version>
    </dependency>
</dependencies>
```

Make sure you update the version number to the current one when you try this.

To conclude this chapter, let's have a look at a simple Sparkplug Java application bundled with Tahu. Named SparkplugListener, it shows how to subscribe to Sparkplug messages on a broker and parse them.

Most of the code contained in the file leverages Eclipse Paho to connect to a broker and subscribe to the spBv1.0/# topic. The messageArrived method that I reproduced contains the Sparkplug-specific logic:

```
@Override
public void messageArrived(String topic, MqttMessage message) throws
Exception {
    Topic sparkplugTopic = TopicUtil.parseTopic(topic);
    ObjectMapper mapper = new ObjectMapper();
    mapper.setSerializationInclusion(Include.NON_NULL);

    System.out.println("Message Arrived on Sparkplug topic " +
    sparkplugTopic.toString());

    SparkplugBPayloadDecoder decoder = new SparkplugBPayloadDecoder();
    SparkplugBPayload inboundPayload = decoder.buildFromByteArray(message.
    getPayload());

    // Convert the message to JSON and print to system.out
    try {
        String payloadString = mapper.writeValueAsString(inboundPayload);
System.out.println(mapper.writerWithDefaultPrettyPrinter().writeVal
ueAsString(inboundPayload));
    } catch (Exception e) {
        e.printStackTrace();
    }
}
```

The critical line of code is this one, where the instance of the `SparkplugBDecoder` class is used to transform the raw bytes from the payload into something that can be converted into JSON by the Jackson `ObjectMapper`:

```
SparkplugBPayload inboundPayload = decoder.buildFromByteArray(message.
getPayload())
```

As you can see, the Tahu library abstracts the complexities of manipulating the Sparkplug payloads. The Tahu repository contains several examples of Sparkplug publishers that you could take inspiration from. Some are stand-alone and simulate devices, while others can run on Raspberry Pis.

CHAPTER 6

DDS

Communiquer, c'est mettre en commun; et mettre en commun, c'est l'acte qui nous constitue.

To communicate is to share, and sharing is the act that constitutes us.

—Albert Jacquard, *Petite philosophie à l'usage des non-philosophes*

Servers are potential points of failure, whatever their workload or physical location is. The Internet and TCP/IP are designed to ensure traffic will reach its destination through alternate routes if specific nodes or larger parts of the network fail. However, the protocols we have explored together up to now all rely on some sort of server. Of course, you can make those servers more resilient by leveraging high-availability features and deploying redundant network connections. But the farther away you are getting from the Cloud or the corporate data center, the harsher the environment will be. For many real-time and mission-critical use cases, having a single point of failure in the architecture is unacceptable, especially since brokers and servers also add latency and jitter to the data flow.

Peer-to-peer networks have for a long time represented a more fault-tolerant alternative to centralized approaches. In such networks, nodes connect directly to as many other nodes as possible. In addition, peer-to-peer networks increase parallelism in the architecture since there is no single bottleneck from a performance standpoint.

In the domain of IoT and edge computing, DDS is without a doubt one of the best-established examples of a protocol relying on a decentralized approach. In this chapter, you will learn about DDS fundamentals and discover Eclipse Cyclone DDS, a portable implementation of the protocol.

F. Desbiens, *Building Enterprise IoT Solutions with Eclipse IoT Technologies*, https://doi.org/10.1007/978-1-4842-8882-5_6

What Is DDS?

The *OMG Data Distribution Service* (DDS) is a protocol and API standard for data-centric connectivity from the Object Management Group (OMG). Development of the DDS specification started in 2001. At the time, the main contributors were Real-Time Innovations (RTI) and Thales Group. The OMG made version 1.0 of the DDS specification available in December 2004. Version 1.1 followed in December 2005 and version 1.2 in December 2006. The most recent version of the core DDS specification is version 1.4, from March 2015.

DDS relies on a data-centric, publish/subscribe approach, sometimes described by the DCPS acronym. As in MQTT, messages are published to topics clients can subscribe to. In DDS, however, entities exist within global data spaces (domains) that can be further divided into partitions.

A significant difference between DDS and other IoT protocols is the importance given to payload definitions. DDS topics are not just mere filters used to determine which subscribers will get a message. They are defined by a name, a type, and a set of QoS policies. Since DDS is language independent, topic types are expressed in various syntaxes, with the *Interface Definition Language* (IDL) being the most common choice.

Note If you are already familiar with IDL, you may wonder how DDS relates to CORBA. The only two things that DDS has in common with CORBA are that it uses a subset of IDL and both are OMG specifications. Also, if you are familiar with CORBA, you are probably as old, if not older than me.

Over time, the DDS specification has been enhanced through the three following companion specifications:

- **DDSI-RTPS:** Defines the *Real-Time Publish-Subscribe Protocol* (RTPS) DDS interoperability wire protocol. The latest version is v2.5. Implementations are identified by an RTPS vendor ID assigned by the OMG.

- **DDS-XTypes:** Defines an extensible type system for DDS Topic Types. The latest version is v1.3.

- **DDS-Security:** Defines a security model and Service Plugin Interface (SPI) architecture that DDS implementations can use. The latest version is v1.1.

The DDS specification and the three companions listed previously are commonly called the DDS core specifications.

The OMG also published extensions to the DDS specification. They introduce a distributed services framework (DDS-RPC), in addition to XML and JSON syntaxes to represent DDS resources. A few gateways are also available, including one supporting the OPC UA protocol. Information about these resources can be found at `www.dds-foundation.org/omg-dds-standard/`.

This chapter will focus exclusively on the latest versions of the core DDS specifications. When selecting a DDS implementation, you should always pay attention to the specifications and extensions they support.

The best way to understand how the core specifications fit together is by looking at the DDS network stack. Please note that work-in-progress initiatives such as adding support for Time-Sensitive Networks (TSN) are not considered.

Protocol Stack

Figure 6-1 illustrates the DDS protocol stack. Please note that DDS requires IP. The core specification mandates support for UDP as a transport, but Cyclone DDS also supports TCP. This makes it a good choice for battery-operated constrained devices and more powerful devices alike. It is worth mentioning that a few vendor-specific extensions enable support for non-IP networks.

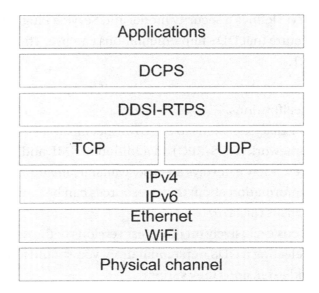

Figure 6-1. *The DDS network stack*

DDSI-RTPS is a wire protocol; the DDSI in the acronym stands for *DDS Interoperability,* while RTPS signifies *Real-Time Publish-Subscribe.* Initially, DDS was an API only, and implementers leveraged several incompatible wire protocols. This led to fragmentation. DDSI-RTPS aimed to foster interoperability between implementations.

Here is a list of the main features of RTPS:

- **Performance and quality of service properties:** Those features enable best-effort and reliable publish/subscribe communications for real-time applications.

- **Fault tolerance:** By design, DDSI-RTPS allows the creation of networks without single points of failure.

- **Extensibility:** The protocol can be extended and enhanced without compromising backward compatibility or interoperability.

- **Plug-and-play connectivity:** Nodes are discovered automatically and can join and leave the network anytime without reconfiguration.

- **Modularity:** Constrained devices can implement a protocol subset and still join the network.

- **Type safety:** Preventing type errors means that the whole infrastructure is more reliable.

The layer above DDSI-RTPS is the DDS API itself, also called *Data-Centric Publish-Subscribe* (DCPS). Its focus is the distribution of data between applications. Its features can be divided into the following three categories:

- **Publishing applications:** DCPS enables you to identify the data objects published by the application and provide values for these objects.

- **Subscribing applications:** DCPS allows you to identify which data objects to subscribe to and access their values.

- **Infrastructure:** DCPS provides everything you need to support publishing and subscribing applications. You will use it to define topics, attach type information to the topics, create publisher and subscriber entities, and attach QoS policies to all these entities.

When DCPS is used, the underlying DDSI-RTPS protocol is not visible to the programmer. Let's now have a deeper look at the features of DCPS, starting with its conceptual model.

Publish and Subscribe

The DDS specification defines several concepts involved in the transmission of information. Sending information requires a *Publisher* and a *DataWriter*; receiving information requires a *Subscriber* and a *DataReader*. *Topics* make it possible to match subscriptions to publications. You can apply distinct QoS properties on the publishing and subscribing sides to control their behavior. Publishers and subscribers can be reused at runtime if desired. Figure 6-2 illustrates how the concepts of DCPS fit together.

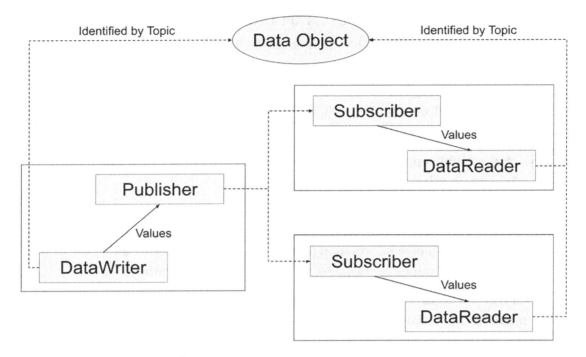

Figure 6-2. *Overview of DCPS concepts*

There are a few nuances that are not apparent in the diagram. The following list provides additional details about DataReaders, DataWriters, Publishers, and Subscribers:

- **A DataWriter is always associated with a single topic.** However, an application can contain several DataWriters sending data to the same topic.

- **Publishers are responsible for data transmission.** Publishers manage and own DataWriters. A single Publisher can own several DataWriters, but DataWriters belong to one Publisher only.

- **A DataReader is always associated with a single topic.** An application can contain several DataReaders reading data from the same topic.

- **Subscribers are responsible for data reception.** Subscribers manage and own DataReaders. A single Subscriber can own several DataReaders, but DataReaders belong to one Subscriber only.

I will describe the specifics of reading and writing data later in this chapter. For the time being, please note that the various DDS specifications do not describe data updates made to a topic as messages but rather as *samples*.

Note   The term *sample* comes from the world of data acquisition. Data acquisition is the process of sampling signals that express real-world physical conditions with sensors and converting the resulting samples into digital values.

DataReaders and DataWriters are strongly typed. The types they use are those defined for the topics. When building a DDS application, you will define topic types yourself using one of the supported syntaxes, with IDL being the most typical choice.

Listing 6-1 shows an IDL type definition for a temperature sensor. You will notice it looks very similar to a C or C++ structure.

Listing 6-1. IDL definition of a temperature sensor

```
// TempSensor.idl
  enum TemperatureScale {
    CELSIUS,
    FAHRENHEIT,
    KELVIN
  };

  @topic struct TempSensorType {
    @key short id;
    float temp;
    TemperatureScale scale;
  };
```

Before using the type in your application, you will need to use the appropriate pre-processor to generate implementation classes. Such pre-processors usually support several programming languages.

In DDS, publication and subscription happen in the context of a fully distributed Global Data Space (GDS). I will now explain what that means.

Global Data Space

DDS applications see all data as local; in other words, when a DDS application reads data, it does so from what looks like a local data store. DDS takes care of data transmission, ensuring subscribers get the data they request. The GDS is also responsible for performing the automatic discovery of publishers and subscribers. There is no central registry of connected clients, like the broker is in MQTT. The advantage of this approach is that clients do not need to be configured before connecting. The auto-discovery mechanisms that are part of DDS take care of everything.

Although the G in GDS means "Global," DDS establishes two ways to scope information access: *domains* and *partitions*. I will now explain what each one entails.

Domains

You can see domains as a virtual network specific to the applications that joined it. Domains cannot communicate with one another. If your application needs to perform operations across multiple domains, you must maintain separate connections to both domains and mediate the data exchanges yourself.

Figure 6-3 shows data writers' and readers' interactions in the context of a single nonpartitioned domain inside the GDS. The diagram highlights that distinct QoS properties can be used on the various participants.

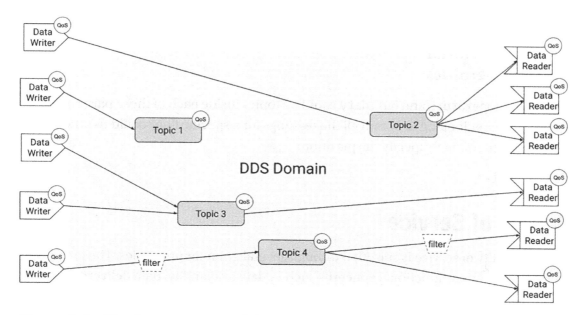

Figure 6-3. *The data-centric model of DDS*

Figure 6-3 also shows that it is possible to apply filters to subscriptions. I will cover this concept later.

Partitions

Partitions are a logical group of topics inside a domain. A name identifies them, and your application needs to join them explicitly before publishing or subscribing to the topics inside. One interesting twist is that applications can join a partition by using its exact name or specifying wildcards. Such wildcards must comply with the POSIX fnmatch API (1003.2-1992 Section B.6).[1] If desired, you can also hierarchically structure partition names by defining a naming convention; the column is a good character to use as a separator in that context.

To make this more concrete, let's go back to our example about museums. In previous chapters, we were deploying infrastructure to monitor temperature and humidity in every room of the *Musée du Louvre*. A possible DDS partition naming convention for this use case could be

```
m-[museum name]:fl-[floor]:rm-[room]
```

---

[1] https://standards.ieee.org/standard/1003_2-1992.html

Actual partition names could look like this:

```
m-Louvre:fl-1:rm-101
m-Louvre:fl-2:rm-205
```

Then, `temperature` and `humidity` could be topics inside each of those partitions.

A subscriber looking to obtain all the readings for a specific floor could use the following expression to specify the partition:

```
m-Louvre:fl-*
```

Quality of Service

In DDS, quality of service is specified through several granular properties. Those properties cover nonfunctional concerns such as data availability, data delivery, timeliness, resource usage, and configuration.

When the QoS for a DataReader and a DataWriter do not match, data will only be transmitted if the QoS requested by the DataReader is not more stringent than the one offered by the DataWriter. This ensures that the guarantees of higher levels of quality of service are preserved.

Let's now review the most used QoS policies offered by DDS.

Data Availability

The policies of the data availability group make it possible to decouple applications across time and space.

The `DURABILITY` policy determines the lifetime of the data written to a domain. It defines the following four levels of durability:

- **VOLATILE:** Data is transmitted to the subscribers that are currently connected. No copy is kept for offline subscribers.

- **TRANSIENT_LOCAL:** Publishers store data locally so that offline subscribers will get a copy of the latest data published once they reconnect to the domain. However, no data will be transmitted if the publisher goes offline.

- **TRANSIENT:** A copy of the latest data published is kept in the GDS outside of the local scope. This ensures offline subscribers will get a copy of the data once they reconnect to the domain. The data will not persist if the system is shut down.

- **PERSISTENT:** A copy of the latest data published is kept in the GDS outside of the local scope and written to persistent storage – typically the filesystem or a database. This ensures offline subscribers will get a copy of the data once they reconnect to the domain. The data will be restored once the system comes back online after a shutdown.

The LIFESPAN QoS policy controls the validity interval of the published data. The default value for this policy is infinite, which means data will not expire. If a value is specified, data elements older than the LIFESPAN threshold will be purged.

The HISTORY QoS policy controls the number of samples (subsequent writes made to the same topic) that readers or writers must store. Possible values are the latest write, the last *n* writes, or all writes.

Data Delivery

The policies of this group influence how published data is made available to subscribers. They directly influence the scalability of a DDS system.

- **PRESENTATION:** This policy affects the ordering and coherency of data updates. The access_scope property determines the granularity of the settings chosen. The supported levels of granularity are INSTANCE, TOPIC, and GROUP. There are two settings in the policy: ordered_access and coherent_access. The former controls whether the order of the updates will be preserved; the latter determines if the grouping of changes specified through the begin_coherent_change and end_coherent_change operations will be maintained. Ultimately, the policy allows reading or writing multiple samples atomically across the scope if coherent_access is enabled and preserving the exact order they were published if ordered_access is enabled. If the value of access_scope is INSTANCE, then this applies to samples with the same key value. If the value is TOPIC, it applies to all the samples published by a specific DataWriter. Finally, if the scope is GROUP, the settings apply to all the DataWriters belonging to the Publisher.

- **RELIABILITY:** This policy controls the level of reliability of data transmission. The possible values are BEST_EFFORT and RELIABLE. The BEST_EFFORT setting provides no delivery guarantees; the Publisher will not attempt to retransmit samples that the Subscriber has not received. When the RELIABLE setting is selected, the Publisher will attempt to retransmit lost samples until received by the Subscriber. In both cases, the ordering of the samples is maintained; data values will not go back from a newer one to an older one.

- **PARTITION:** This policy specifies the association between an instance of a Publisher or Subscriber and partitions. This relies on the name of the partition, which can contain wildcards. When matching Publishers and Subscribers, DDS considers both the topic and partitions.

- **DESTINATION_ORDER:** This policy defines if the samples will be ordered by referencing the time of their reception (BY_RECEPTION_TIMESTAMP), which is the default value, or a timestamp set at the source (BY_SOURCE_TIMESTAMP). The BY_SOURCE_TIMESTAMP setting is needed for maintaining eventual consistency between samples when there are multiple publishers for the same data or when subscriptions to non-VOLATILE data are taken after publishers have begun publishing data.

- **OWNERSHIP:** This policy determines if a specific instance of a Data Object, determined by a topic and a key, will accept concurrent updates by multiple DataWriters or not. If set to SHARED, then multiple DataWriters can write. If set to EXCLUSIVE, then a single DataWriter "owns" the Data Object at any given time. The OWNERSHIP_STRENGTH policy determines which of the DataWriters will obtain write access. The DDS specification does not require implementations to notify DataWriters that they do not own the Data Object they are trying to update.

Data Timeliness

The policies in this group influence the temporal aspects of transmission and can influence ownership of data objects when the OWNERSHIP policy is EXCLUSIVE.

- **DEADLINE:** This policy is useful when a topic is supposed to be updated regularly. The policy defines a contract that the application must fulfill on the publishing side. On the subscribing side, it defines a minimum requirement for publishers that supply data values. DDS applications can be notified of missed deadlines.

- **LATENCY_BUDGET:** This policy provides a way for applications to inform DDS of the urgency of the transmitted data. The DDS specification states that the value is a hint left to the interpretation of the implementation. Some implementations will treat it as a delivery window for data transmission.

- **TRANSPORT_PRIORITY:** This policy allows applications to indicate the relative priority of the data associated with a topic. The value is a 32-bit signed integer; higher values indicate higher priority. The DDS specification states that this policy is a hint, and the actual runtime behavior is implementation dependent.

Resources

The policies in this group enable you to define limits to the resources consumed by the DDS system. This makes it possible to keep the infrastructure responsive even in cases where bandwidth and storage are limited.

- **TIME_BASED_FILTER:** This policy controls the sampling rate of a DataReader. It is useful when subscribers lack the network bandwidth, memory, or processing power required to process data over a certain rate. Specifically, the policy enables you to define a minimum_separation period between two samples on a topic. In other words, this policy acts as a rate-limiting mechanism.

- **RESOURCE_LIMITS:** This policy allows applications to define the maximum number of samples and samples per instance. When applicable, this effectively defines the storage available to hold topic instances and their related historical samples.

133

Configuration

This group's policies focus on the definition and transmission of user-defined values. Such values are especially useful in the bootstrapping process.

- **USER_DATA:** This policy allows applications to associate metadata to domain participants, DataReaders, and DataWriters. This metadata is then distributed through a system topic. The policy represents a generic extensibility mechanism and is often used to provision security credentials.

- **TOPIC_DATA:** This policy allows applications to associate metadata to a topic. This metadata is then distributed through a system topic. The policy represents a generic extensibility mechanism.

- **GROUP_DATA:** This policy allows applications to associate metadata to Publishers and Subscribers. This metadata is then distributed through system topics. Applications can leverage such metadata to get additional control over subscription matching.

Topics

As you realize now, DDS QoS policies are much more granular than QoS levels in MQTT. Topics represent another difference between the two protocols. As I mentioned earlier, DDS topics are defined by a name, a type, and a set of QoS policies. We will now explore what this means together.

Topic Types

Although alternatives are available, IDL remains the most common choice to define DDS topic types. When expressed in IDL, topic types are composed of a key and an IDL `struct`. The `struct` can contain an arbitrary number of fields. Those fields can be either a primitive type, a template type, or a constructed type.

Table 6-1 shows the IDL primitive types available when defining DDS topics.

Table 6-1. *IDL primitive types used to define DDS topics*

Primitive Type	Size (bits)
boolean	8
octet	8
char	8
short	16
unsigned short	16
long	32
unsigned long	32
long long	64
unsigned long long	64
float	32
double	64

As you can see, there is no int type in IDL. You can use short, long, or long long instead; simply pick the appropriate size for the data you wish to transmit.

There are two template types available. One can define strings of defined or undefined (unbounded) length. The other is a random-access ordered container called *sequence*, conceptually similar to a vector in most high-level programming languages. Table 6-2 provides more details.

Table 6-2. *IDL template types used to define DDS topics*

Template Format	Example Declarations
String<length = UNBOUNDED$>	string museumName; string<32 museumShortName;
sequence<T,length = UNBOUNDED>	sequence<octet> floorRooms sequence<octet, 5> lastFive sequence<TempSensorType> tSens sequence<TempSensorType, $5>$ tS

Finally, you can use three distinct IDL constructed types in your DDS types: enum, struct, and union. IDL enums are simply enumerations of possible predefined values. IDL structs are equivalent to the same construct in the C language. As for IDL unions, they define a structure that can contain only one of several alternative members. Table 6-3 contains examples of the three allowed constructed types.

Table 6-3. *IDL constructed types used to define DDS topics*

Constructed Type	Example Declarations

enum	```enum TemperatureScale {```
	 ``` CELSIUS,```
	 ``` FAHRENHEIT,```
	 ``` KELVIN```
	 ```};```
	 ```enum SensorType {SIMPLE, TEMP};```

struct	```struct Sensor { short id;};```
	 ```struct TempSensor {```
	 ``` short id;```
	 ``` TemperatureScale scale;```
	 ```};```

union	```union RoomSensors switch(SensorType){```
	 ``` case SIMPLE: Sensor simpleSensor;```
	 ``` case TEMP: TempSensor tSensor;```
	 ```};```

In the end, any DDS topic type is simply a struct that contains nested primitive types, template types, and constructed types as described previously.

The DDS-XTypes specification brought additional flexibility in this domain. It introduced single inheritance and optional fields, for example. In addition, it added tolerance for slight variations between types on the DataReader and DataWriter sides. This makes it possible for you to evolve the various components in a system without updating everything simultaneously.

Topic Keys, Instances, and Samples

When defining a DDS topic type, you will need to decide if you wish the type to have a key. Topics that are keyless will only have one instance; in other words, they will be singletons. On the other hand, keyed topics will have as many instances as key values at runtime.

In Listing 6-1, I defined a temperature sensor that contained a short value named id. The @key annotation is used to denote the components of the key:

```
@key long userID;
```

Topic keys can be a primitive type, an enumeration (enum), or a string nested at any level of the type declaration. For example, if I wished to use both the id and scale (type TemperatureScale, which is an enum) as the topic key, I would annotate both properties with @key:

```
@topic struct TempSensorType {
  @key short id;
  float temp;
  @key TemperatureScale scale;
};
```

To make the DDS type keyless, all you need to do is omit the @key annotation.

DDS Topic types are comparable to tables in a relational database and DDS Topic instances to rows. DDS Topic types also map to classes in object-oriented languages and DDS Topic instances to object instances.

Filtering

Frequently, a subscriber will care only about a subset of the samples published to a topic. A typical use case would be to take specific actions once the values are outside a specific range. DDS offers content filtering features that allow subscribers to get only the relevant samples. The DDS specification describes this as content-based subscriptions.

In DDS, filtering involves creating a content-filtered topic by applying a filter to an existing topic. The filter is specified through an expression using one or several supported filtering operators. Table 6-4 lists the filtering operators supported by DDS.

Table 6-4. *Supported operators for DDS filters and query conditions*

Filter Operator	Meaning
=	Equal
	Not equal
>	Greater than
<	Less than
>=	Greater than or equal
<=	Less than or equal
BETWEEN	Between and inclusive range
LIKE	Matches a string pattern

Filter expressions can also contain logical operators such as AND, NOT, and OR.

Suppose I wish to subscribe to a filtered topic to be notified when the temperature is outside the acceptable range in a room inside the *Musée du Louvre*. The type of the topic is TempSensorType, as defined in Listing 6-1. The filter expression could look like this:

```
temp NOT BETWEEN 22.5 AND 23
```

Most DDS implementations make it possible to create the filter expression as a parameterized string to which actual values are passed later.

Reading and Writing Data

Filtering is an effective strategy to minimize bandwidth and memory consumption by transmitting only the samples it cares about to a subscriber. However, it is also possible for application code to select only a subset of the samples present in the cache of a DataReader to process them. To understand how this works, you first need to learn more about how data is written in DDS.

Writing

Basic data writing in DDS is simple: just call the `write` function or method on a DataWriter. Behind the scenes, the life cycle of topic instances impacts resource consumption on the local node and remote ones.

There are three states in the life cycle of a topic instance. The following list describes each of the three states:

- **ALIVE:** At least one DataWriter has registered this instance of the topic. This means the DataWriter intends to provide data updates and has reserved resources related to the instance.

- **NOT_ALIVE_NO_WRITERS:** There are no more DataWriters who have registered the topic instance. All DataWriter resources related to the instance have been released. DataReaders no longer expect new samples. DataWriters need to unregister topic instances when they don't need them anymore because this causes resource leaks on the publishers and all related subscribers. Topic instances for an application that crashed will typically be in the NOT_ALIVE_NO_ WRITERS state but could go back to ALIVE once the application comes back online.

- **NOT_ALIVE_DISPOSED:** This state indicates that the topic instance needs to be discarded. This is the state an application should set its topic instances to in the case of a graceful termination or, for example, when an object described by the instance no longer exists.

DDS provides a form of automatic life cycle management for topic instances. By default, when a DataWriter object is destroyed, DDS will set topic instances to the NOT_ALIVE_DISPOSED state if no other active DataWriters have registered against the topic instance. The behavior can usually be overridden through a configuration setting, causing DDS to place the instances in the NOT_ALIVE_NO_WRITERS state instead.

As a developer, you can explicitly manage the life cycle of topic instances by using the APIs exposed by DataWriter. This will give you full control over topic instance registration and disposal. Usually, this approach results in finer-grained management of system resources. Registration results in resource allocations; those resources are freed when the DataWriter unregisters.

Note   In the case of keyless topics, the topic instance is a singleton. This essentially means that the life cycle of the topic is the same as the life cycle of the DataWriter.

Reading

DataReaders expose two operations that enable you to access data: *read* and *take*. Both operations will give you a copy of the data. But read will leave the data in the DataReader's cache, while take will remove it from the cache. Using read is akin to having a retained message set on a topic in MQTT. Which one to use depends on your use case. Read is useful when your DDS topics represent distributed states and take when they represent distributed events.

Each sample received by a DataReader has a companion SampleInfo describing the sample's properties. Here is a list of the most important properties:

- **sample_state:** The sample state can be READ or NOT_READ. READ is, of course, only possible if data was obtained through a read, not a take.

- **instance_state:** This indicates the status of the topic instance. Possible values are ALIVE, NOT_ALIVE_NO_WRITERS, and NOT_ALIVE_DISPOSED.

- **view_state:** The value of this property will be NEW if the sample is the first one received on the topic instance and NOT_NEW otherwise.

- **disposed_generation_count and no_writers_generation_count:** SampleInfo features counters that record the number of times the state of a topic instance changed. For example, you could use them to know how many times a topic instance was in the ALIVE state after being in the NOT_ALIVE_DISPOSED state.

- **source_timestamp:** The timestamp is important since data timeliness QoS policies rely on it. It is also often used by applications to represent the time an observation was made.

- **valid_data:** This flag indicates whether the sample contains data or not. Some samples do not contain data since they are sent to convey a change in the value for instance_state.

DDS allows you to select a subset of the data at reading time, regardless of whether you use read or take. You can base data selection on *state* or *content*. State-based selection involves the view_state, instance_state, and sample_state properties of SampleInfo. You could, for example, read data provided by instances that just showed up in the GDS by issuing a read for samples where view_state is NEW, sample_state is NOT_READ, and instance_state is ALIVE. Content-based selection, on the other hand, involves the actual sample data. DDS provides content-based selection through queries, which use the same syntax as DDS filters. Table 6-4 lists the supported operators.

Conceptually, filters and queries are close to each other. How do they differ, then? Filtering controls the data received by the data reader: the data that does not match the filter is not inserted into the DataReader cache. On the other hand, queries are about selecting the transmitted data already found in the DataReader cache.

DDS applications can poll DataReaders periodically to check if samples have been published. In most cases, this approach will waste CPU time and reduce battery life in the case of constrained devices. DDS offers two alternatives to polling for checking data availability: *listeners* and *waitsets*. You can register listeners against DataReaders to be notified of data availability or violations of some QoS policies. As for waitsets, they provide a more generic mechanism to execute logic when specific conditions are met. They make it possible to bind callback functions or methods to various events.

Security

The DDS specification itself says little about security, only mentioning that the USER_DATA QoS policy can be used to transmit security tokens. This is perhaps a bit naive, given that such a token would be transmitted in cleartext. The DDS Security specification, on the other hand, defines the *Security Model and Service Plugin Interface* (SPI) architecture. DDS implementations can enforce the security model by calling those SPIs. The specification also defines a set of built-in implementations of the SPIs.

The SPI architecture is generic and does not mandate specific security technologies. The exact details of the security functions and the implementation of the different plugins are specific to the DDS implementation used and the application.

The core of the DDS Security architecture consists of five plugins:

- **Authentication Service plugin:** Provides one-way and mutual authentication algorithms.

- **Access Control Service plugin:** Handles authorization on publish/subscribe operations as well as implementation-specific operations.

- **Cryptographic Service plugin:** Provides key generation and exchange interfaces, encryption, hashing, and digital signatures. It also handles message authentication codes.

- **Logging Service plugin:** Implements event logging. It makes it possible to audit all DDS Security-related events.

- **Data Tagging plugin:** Provides a way to add tags to samples. From a security perspective, this means you can attach security labels to the data.

Of those, the Logging Service and Data Tagging plugins are optional.

Here is a high-level overview of the security flows enabled by the plugins. Every entity joining a DDS domain must authenticate; they get a security token in return. The infrastructure can generate a shared secret (key) during the authentication process, which the Cryptographic Service plugin can use to create and distribute derived cryptographic keys to other entities. For example, this can support the establishment of TLS secure communication channels. Moreover, the Cryptographic Service plugin provides an API that applications can leverage to enforce message confidentiality, integrity, authenticity, and nonrepudiation. The Access Control plugin, on the other hand, ensures only authorized entities can publish or subscribe to specific topics. It relies on the token obtained at authentication time to establish the identity of an entity. The automatic discovery process built into DDS can also be protected through the Access Control Service plugin.

Most DDS implementations can integrate with various authentication servers, use OAuth tokens or equivalents, and encrypt traffic using TLS (TCP) or DTLS (UDP).

Eclipse Cyclone DDS

Eclipse Cyclone DDS is a mature and fully featured implementation of DDS hosted at the Eclipse Foundation. It has been used in several commercial implementations already. The project focuses on compliance with all the core DDS specifications, namely:

- Data Distribution Service (DDS)

- DDS Interoperability Wire Protocol (DDSI-RTPS)

- DDS Security (DDS-SECURITY)

- Extensible and Dynamic Topic Types for DDS (DDS-XTypes)

The core Cyclone DDS library is written in C. The Cyclone DDS project team also maintains bindings in Python and C++ (ISO/IEC C++ PSM). Cyclone DDS is made available under the Eclipse Distribution License v1.0 and the Eclipse Public License v2.0.

The official web resources for Cyclone DDS are as follows:

- **Website:** `https://cyclonedds.io`

- **Eclipse project page:** `https://projects.eclipse.org/projects/iot.cyclonedds`

- **Code repository:** `https://github.com/eclipse-cyclonedds/cyclonedds`

At the time of writing, the current version was 0.8.2, released on January 10, 2022. One of the 0.8.x series innovations was to support a zero-copy, shared memory approach through integration with Eclipse iceoryx.

Installation

At the time of writing, the Cyclone DDS project did not publish packages for its binaries. Consequently, you will need to compile it yourself. Fortunately, the process is straightforward.

Prerequisites

It is possible to build Cyclone DDS on Linux, macOS, and Windows. With some caveats, it will also build on *BSD operating systems. You will need to install the following software packages:

- **C compiler:** This is typically GCC on Linux and the Microsoft Visual C++ (MSVC) compiler on Windows. The LLVM compiler shipping with the XCode Command Line Tools is a typical choice on macOS.

- **GIT:** You will need the GIT version control system to clone the Cyclone DDS repository.

- **CMake:** Cyclone DDS leverages CMake for its build system. You will need CMake version 3.10 or later.

- **OpenSSL:** You need OpenSSL to support TLS communications. Ideally, you should install OpenSSL v1.1 or later. On some systems, you will also need to install libssl. You can install the latter by executing `apt-get install libssl-dev` on Debian and its derivatives, and `yum install openssl-devel` on Fedora and its derivatives.

- **Bison:** GNU Bison is a general-purpose parser generator. It is used to parse IDL type definitions in Cyclone DDS.

Cloning the Repository

Cyclone DDS is available on GitHub. Cloning the repository and getting ready to build the source is straightforward. Here are the commands you should execute in the directory of your choice:

```
git clone https://github.com/eclipse-cyclonedds/cyclonedds.git
cd cyclonedds
mkdir build
```

Compiling the Code

On Linux and macOS, the only parameter you need to pass to the build system is the directory where you wish to install the binaries. In my case, I chose /opt/cyclone. Here are the commands you need to run to configure the build:

```
cd build
cmake -DCMAKE_INSTALL_PREFIX=/opt/cyclone -DBUILD_EXAMPLES=ON ..
```

If all the prerequisites are installed correctly, you should see output like this:

```
-- The C compiler identification is GNU 11.2.0
-- Detecting C compiler ABI info
-- Detecting C compiler ABI info - done
-- Check for working C compiler: /usr/bin/cc - skipped
-- Detecting C compile features
-- Detecting C compile features - done
...
-- Found BISON: /usr/bin/bison (found suitable version "3.7.6", minimum
required is "3.0.4")
-- Configuring done
-- Generating done
-- Build files have been written to: /home/fdesbiens/cyclonedds/build
```

If all goes well, you can compile the code using this command:

```
cmake --build .
```

You should see output like this:

```
Scanning dependencies of target ddsrt-internal
[  0%] Building C object src/ddsrt/CMakeFiles/ddsrt-internal.dir/src/
atomics.c.o
[  1%] Building C object src/ddsrt/CMakeFiles/ddsrt-internal.dir/
src/avl.c.o
[  1%] Building C object src/ddsrt/CMakeFiles/ddsrt-internal.dir/src/
bswap.c.o
[  2%] Building C object src/ddsrt/CMakeFiles/ddsrt-internal.dir/src/io.c.o
...
Scanning dependencies of target schema
[100%] Generating cyclonedds.rnc, cyclonedds.xsd, manual/options.md
[100%] Built target schema
```

If there are no compilation errors, congratulations! You are now ready to build your own DDS applications. The only step left is to copy the binaries in the specified folder. To do this, just run the following command:

```
cmake --build . --target install
```

You should prefix the command with sudo if the installation folder requires administrative privileges.

Testing Your Setup

Since I used the -DBUILD_EXAMPLES=ON switch when configuring the build, I now have access to the binaries of the Cyclone DDS samples in the bin subdirectory. In other words, if the build directory was /home/fdesbiens/cyclonedds/build, then the binaries for the samples are in /home/fdesbiens/cyclonedds/build/bin after the build.

The roundtrip sample application allows you to measure the performance of your environment by sending and receiving a single message. The application is made of two executables: RoundtripPong and RoundtripPing. To run the sample, open a command line and start the Pong component:

```
./RoundtripPong
```

You should see this output:

```
Waiting for samples from ping to send back...
```

Open a second command line and start the Ping component:

```
./RoundtripPing 0 0 0
```

This will start the sample in a mode where samples with an empty payload will be sent indefinitely. The first number controls the payload size, the second specifies the number of samples sent, and the third defines a timeout in seconds. Executing Ping will result in output resembling this:

```
# payloadSize: 0 | numSamples: 0 | timeOut: 0

# Waiting for startup jitter to stabilise
# Warm up complete.

# Latency measurements (in us)
...
```

You should see information about the latency, write access time, and read access time displayed after a few seconds.

> **Note** For a more precise assessment of the performance of your DDS environment, you could use the ddsperf tool built by the Cyclone DDS team. The project repository contains a series of scripts the project team uses to benchmark the library.

The Hello World! Sample

With the libraries compiled and installed, you are now ready to work on your applications. To explain the basics of writing a publisher and subscriber in C using Cyclone DDS, I will now explore with you the code of the "Hello World!" Cyclone DDS sample application. The Cyclone DDS website[2] explains how to build the "Hello World!" example in detail.

The "Hello World" application comprises a subscriber and a publisher. The publisher sends a single sample on a topic, which is then picked by the running subscribers, if any.

Data Model

The topic type used in "Hello World!" is named Msg. It is defined in the HelloWorldData.idl file, reproduced in Listing 6-2.

Listing 6-2. HelloWorldData.idl

```
module HelloWorldData
{
  @topic struct Msg
  {
    @key long userID;
    string message;
  };
};
```

The module declaration at the top is akin to a scope or namespace. The numerical userID defined as one of the two properties is declared as the key for type Msg.

---

[2] https://cyclonedds.io/docs/cyclonedds/latest/GettingStartedGuide/helloworld.html

During the compilation process, the IDL will be translated into a C `struct` that looks like this:

```
typedef struct HelloWorldData_Msg
{
  int32_t userID;
  char * message;
} HelloWorldData_Msg;
```

Publishing

The "Hello World!" subscriber code is contained in subscriber.c. The file contains a single main method. Variable declarations can be found at the top. Here are the main ones:

```
  dds_entity_t participant;
  dds_entity_t topic;
  dds_entity_t reader;
  HelloWorldData_Msg *msg;
  void *samples[MAX_SAMPLES];
  dds_sample_info_t infos[MAX_SAMPLES];
  dds_return_t rc;
  dds_qos_t *qos;
```

The first three variables of type `dds_entity_t` are needed to create a DataReader. The next three are buffers to hold the message, the message's sample, and the `sample_info`. `MAX_SAMPLES` is a constant, and its value is 1. Finally, `dds_return_t` is the return value for the read operation, while `dds_qos_t` specifies the QoS.

To create a data reader, you need a domain participant, a topic, and defined QoS properties. The following is the code used to achieve this, with error handling removed:

```
participant = dds_create_participant (DDS_DOMAIN_DEFAULT, NULL, NULL);
topic = dds_create_topic (
    participant, &HelloWorldData_Msg_desc, "HelloWorldData_Msg",
    NULL, NULL);
qos = dds_create_qos ();
dds_qset_reliability (qos, DDS_RELIABILITY_RELIABLE, DDS_SECS (10));
reader = dds_create_reader (participant, topic, qos, NULL);
```

The main parameter needed to create the topic is the name of the DDS type (HelloWorldData_Msg). As for the QoS, it is set for RELIABLE delivery and a maximum blocking time of ten seconds. If no timeout is desired, it is possible to use DDS_INFINITY as the value.

We now have a DataReader. Time to read published values, if any, by calling dds_read.

```
rc = dds_read (reader, samples, infos, MAX_SAMPLES, MAX_SAMPLES);
```

The read is performed in a while(true) loop, but the break instruction in the following code ensures execution will continue when a message is read. The program prints the userID and message from the first sample received.

```
if ((rc > 0) && (infos[0].valid_data))
{
  /* Print Message. */
  msg = (HelloWorldData_Msg*) samples[0];
  printf ("=== [Subscriber] Received : ");
  printf ("Message (%"PRId32", %s)\n", msg->userID, msg->message);
  fflush (stdout);
  break;
}
else
{
  /* Polling sleep. */
  dds_sleepfor (DDS_MSECS (20));
}
```

I highlighted how important it is to release resources that are no longer needed when working in DDS. Hoarding DataReaders that should be disposed of leaks local and remote memory. In this case, the application starts by calling HelloWorldData_Msg_free. This is a generated function tailored to the DDS type used. Then, the application calls dds_delete on the domain participant. Doing this deletes recursively everything it contains. In this case, this means the topic instance and DataReader we just created.

```
HelloWorldData_Msg_free (samples[0], DDS_FREE_ALL);
rc = dds_delete (participant);
```

One important thing to note about the "Hello World!" sample application is that it relies on a polling approach. It calls `dds_read` in a loop and will sleep for 20 milliseconds if no messages samples are available (`dds_sleepfor`). On the other hand, the `roundtrip` sample takes a different approach and leverages listeners and waitsets. You can see this in action in the code for RoundtripPong.

Publishing

The code for the "Hello World!" publisher is contained in publisher.c. The file contains a single main method. Variable declarations can be found at the top. Here are the main ones:

```
dds_entity_t participant;
dds_entity_t topic;
dds_entity_t writer;
dds_return_t rc;
HelloWorldData_Msg msg;
uint32_t status = 0;
```

As you can see, they are remarkably like the ones used in the subscriber. The main difference lies in the `status` variable, which is an unsigned integer that will contain the current set of statuses of the DataWriter.

Before creating the DataWriter, you need to create a domain participant and topic as shown in the following:

```
participant = dds_create_participant (DDS_DOMAIN_DEFAULT, NULL, NULL);
topic = dds_create_topic (
  participant, &HelloWorldData_Msg_desc, "HelloWorldData_Msg", NULL, NULL);
writer = dds_create_writer (participant, topic, NULL, NULL);
```

Sending out data when no one is listening is wasteful. DDS DataWriters know if there are matching DataReaders, that is, DataReaders defined on the same topic instance somewhere in the domain. Consequently, the application will loop and sleep until the DataWriter has been informed of a publication match. The following is a simplified version of the logic with error handling removed:

```
rc = dds_set_status_mask(writer, DDS_PUBLICATION_MATCHED_STATUS);
...
```

```
while(!(status & DDS_PUBLICATION_MATCHED_STATUS))
  {
    rc = dds_get_status_changes (writer, &status);
    ...
    /* Polling sleep. */
    dds_sleepfor (DDS_MSECS (20));
  }
```

Building and sending the message is straightforward. Simply set the userID and message on the msg variable and call dds_write.

```
msg.userID = 1;
msg.message = "Hello World";

rc = dds_write (writer, &msg);
```

Once this is done, the only thing left is to clean up. Calling dds_delete on the domain participant will free up the DataReader and topic instance resources.

```
rc = dds_delete (participant);
```

As you can see, the code for the publisher and subscriber is very close. The QoS settings are found in the subscriber but not in the publisher. Why is that? The explanation is that the default QoS policies applied by Cyclone DDS to DataWriters were enough to match the DataReader in the subscriber.

Running the Sample

Open a command line and go to the bin subdirectory in the build folder you created to compile Cyclone DDS. Then, execute the subscriber like this:

```
./HelloworldSubscriber
```

You will see this output:

```
=== [Subscriber] Waiting for a sample ...
```

Then, open another command line and execute this command to launch the publisher:

```
./HelloworldPublisher
```

You will see this output:

```
=== [Publisher]  Waiting for a reader to be discovered ...
=== [Publisher]  Writing : Message (1, Hello World)
```

If you go back to the command line where the subscriber is running, you will see this:

```
=== [Subscriber] Received : Message (1, Hello World)
```

One notable aspect of this demo and the `roundtrip` sample is that you did not need to configure anything for things to work. There were no IP addresses or hostnames involved. This is the magic of DDS auto-discovery at work.

Python Version

I mentioned earlier that the Cyclone DDS team maintains Python and C++ bindings for the library. The repositories for those bindings also contain example applications. To conclude this section, let's now look at the Python version of the "Hello World!" example application.[3]

To use the Python binding, also called `cyclonedds-python`, you first need to install Cyclone DDS in your environment, as described earlier in this chapter. Then, follow the installation instructions found in the `README.MD` in the binding's GitHub repository.[4] It is highly recommended to use a virtual environment, poetry, pipenv, or pyenv when working with `cyclonedds-python`.

At the time of writing, OMG provided no official mapping between IDL and Python. The Cyclone DDS team built tools to address that gap; unfortunately, those are specific to `cyclonedds-python` and do not apply to other DDS implementations or bindings. The documentation for `cyclonedds-python` provides more details.[5]

`Cyclonedds-python` includes a command-line IDL compiler that will generate a Python module containing your types from an IDL file. Alternatively, the binding provides IDL data types in the `cyclonedds.idl` package. This is the approach used in the Python version of the "Hello World!" example. You will find the declaration for the `Msg` type at the top of the `helloworld.py` file.

---

[3] https://github.com/eclipse-cyclonedds/cyclonedds-python/blob/master/examples/
helloworld/helloworld.py

[4] https://github.com/eclipse-cyclonedds/cyclonedds-python/

[5] https://github.com/eclipse-cyclonedds/cyclonedds-python/

```
@dataclass
class HelloWorld(IdlStruct, typename="HelloWorld.Msg"):
    data: str
```

The basic steps to publish a sample are the same as in the C version of the example. As shown in the following, you need to create a domain participant, topic, and DataWriter.

```
dp = DomainParticipant()
tp = Topic(dp, "Hello", HelloWorld)
dw = DataWriter(dp, tp)
```

Publishing the sample is straightforward. All you need to do is pass an instance of the HelloWorld object to the DataWriter's `write` method.

```
sample = HelloWorld(data='Hello, World!')
dw.write(sample)
```

Reading data is also very simple. All you need is a DataReader instance. In this case, the code only prints the `data` member from the first sample returned by the `read` method.

```
dr = DataReader(dp, tp)
sample = dr.read()[0]
print(sample.data)
```

And that's it! I am sure you can appreciate the simplicity and readability compared to the C version. For the time being, you can only install `cyclonedds-python` from the source, but the team aims to publish packages to the `pip` repository in the future. Enabling developers to install from source using PyPI is also in the plans.

Cyclone DDS and Robotics

Cyclone DDS has seen tremendous adoption since the inception of the project. Robotics is one domain where the project made a name for itself. At a high level, using DDS with robots makes a lot of sense. After all, they often operate close to one another in the same physical space; making them part of a mesh makes sense. The emergence of autonomous robots and machines increases the need to securely transmit large amounts of data in real time. Given this, it is not surprising to see that Cyclone DDS significantly impacted the ROS 2 ecosystem.

ROS (Robot Operating System) is an open source software development kit for robotics applications. The most recent version, ROS 2, is supported on Linux, Windows, and macOS. Additionally, the micro-ROS project targets embedded platforms running an RTOS. Open Robotics and the ROS 2 Technical Steering Committee members maintain the core ROS source code.

The ROS client library defines an API exposing communication concepts like publish/subscribe to developers. In ROS 2, such features have been built on top of DDS. Or rather, the ROS 2 team built an abstract middleware interface[6] between the client library and specific DDS implementations. This makes it possible for developers to pick the implementation that fits their requirements. The official ROS 2 documentation describes how to work with middleware implementations[7] in greater detail.

In December 2020, the ROS 2 TSC selected Cyclone DDS as the default middleware implementation[8] for the Galactic Geochelone release, made available in May 2021.

---

[6] https://design.ros2.org/articles/ros_middleware_interface.html
[7] https://docs.ros.org/en/galactic/Concepts/About-Middleware-Implementations.html
[8] https://discourse.ros.org/t/ros-2-galactic-default-middleware-announced/18064

CHAPTER 7

zenoh

On aimerait tant pouvoir inventer le dragon qui fera de nous des princes charmants.

We would so love to be able to invent the dragon that will make us prince charming.

—Bernard Arcand, *De nouveaux lieux communs*

It happens to the best of us. I open an old document; I browse code I wrote a few months or a few years ago. At first, there is only a vague feeling of uneasiness. Then, some head scratches. And finally, a murmur or even a shout: "What was I thinking?" The passage of time is often cruel to our creations. This is why I probably did so much refactoring and rewriting throughout my career.

DDS is incontestably flexible and powerful. However, it is also complex and has a steep learning curve. Figuring out how to match QoS policies can be a challenge. Moreover, routing DDS traffic over the public Internet can be hard. And ultimately, the protocol was not designed to address the concerns of constrained devices. Automated discovery is wonderful, but in DDS's case, it generates a lot of network traffic, significantly reducing the battery life of devices deployed in the field. DDS was first standardized in 2003, and it shows.

The Eclipse zenoh protocol provides location-transparent primitives for data in motion through a publish/subscribe model and for data at rest by supporting geographically distributed storage. It is intentionally designed to scale from microcontrollers up to the Cloud. In a way, zenoh takes the best ideas of DDS, adds its own, and removes the complexity to deliver a protocol built from the ground up for IoT and edge computing use cases.

Compared to other protocols covered in this book, zenoh is not described in a formal specification. The project team implemented the core library in Rust, which makes the protocol safer in terms of memory management since the language does not allow

155

© Frédéric Desbiens 2023
F. Desbiens, *Building Enterprise IoT Solutions with Eclipse IoT Technologies*,
https://doi.org/10.1007/978-1-4842-8882-5_7

dangling pointers or null pointers. The core zenoh implementation is accompanied by a growing number of language bindings, of which the C and Python ones are the most mature. The project's website provides full documentation for the APIs.[1]

zenoh Basics

The most important thing to remember about zenoh is that its scope is broader than most other IoT protocols. This becomes apparent when considering the data life cycle, as shown in Figure 7-1. If you think about MQTT, another popular publish/subscribe protocol, you will notice that it focuses strictly on transmission – which is step 2 in the figure. Even when it stores data for offline subscribers, an MQTT broker always does so in a transient way. Zenoh is different: it also supports data at rest and computations – steps 3 and 4 in the figure.

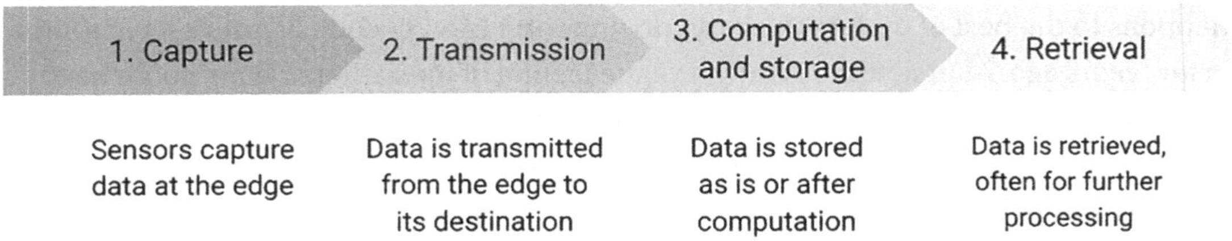

1. Capture	2. Transmission	3. Computation and storage	4. Retrieval
Sensors capture data at the edge	Data is transmitted from the edge to its destination	Data is stored as is or after computation	Data is retrieved, often for further processing

Figure 7-1. *The life cycle of data*

In addition to plain subscriptions, there are two ways to retrieve and compute data in zenoh. One is distributed queries, where various nodes in the zenoh infrastructure will provide part of the result set. The other is distributed computed values, where one or several computation functions are triggered to return a result.

Data retrieval implies data storage. Zenoh supports this through a specialized plugin that I will cover later. For the time being, I will simply say that zenoh enables you to define storage units anywhere in the infrastructure. Those storage units can reside in databases (both relational and nonrelational), memory, or the filesystem of a node.

---

[1] https://zenoh.io/docs/apis/apis/

Key Abstractions

The key/value paradigm is at the core of zenoh. I will now explain how it is implemented and describe a few related abstractions.

Zenoh *keys* are built like filesystem paths on UNIX-like operating systems and are comparable to MQTT topics. They are hierarchical strings where the levels are separated by the forward-slash character ("/"). The following are a few examples of zenoh keys:

```
/museums/Louvre/floors/1/rooms/101/sensors/temperature
/museums/Louvre/floors/2/rooms/202/sensors/humidity
```

The preceding keys point to specific *resources*; they are also called *named data* items in zenoh parlance. It is possible to build expressions involving a set of keys. Such *key expressions* can contain wildcards. The single asterisk ("*") is a single-level wildcard, while the double asterisk ("**") is a multilevel one. For example, the following key expression could be used to subscribe to all the sensors in room 101:

```
/museums/Louvre/floors/1/rooms/101/sensors/*
```

This other expression could return data for all the temperature sensors, independently of their location:

```
/museums/Louvre/**/temperature
```

You will use key expressions in the following situations:

- Subscribing to data

- Creating a storage instance

- Registering a distributed computation (eval)

zenoh also possesses *selectors*. Those are key expressions identifying a set of resources. The zenoh documentation describes the structure of selectors like this:

```
/s1/s2/.../sn?x>1&y<2&...&z=4(p1=v1;p2=v2;...;pn=vn)[#a;b;x;y;...;z]
|           | |              | |                   | |             |
|-- expr ---| |--- filter --| |---- properties ---| |--fragment -|
```

Let's now look at each of the components of the selector structure. Please remember that each of the parts following the path expression itself is optional.

- **expr:** A standard zenoh key expression.

- **filter:** A list of predicates separated by ampersands ("&"). The filter will be applied to the values matching the path expression. Each predicate has the form "field-operator-fieldvalue." The supported operators are <, >, <=, >=, =, and !=.

- **properties**: A string representing key/value pairs separated by a semicolon.

- **fragment:** A list of fields names allowing to return a subset of each value. This works only when a self-describing encoding such as JSON or XML is used.

At the time of writing, the zenoh team had documented but not implemented filters and fragments. In a real application, a selector could look like this:

```
/museums/**/sensors/temperature?celcius>23.5
```

You will use selectors to run distributed queries.

The results of distributed queries can contain values fetched from storage instances or produced by a computation (eval) if the key path used at their creation matches the selector used. Therefore, storages and evals are sometimes described as *queryables* in the zenoh documentation.

As in MQTT, zenoh values are transmitted as a stream of bytes. Values are not strongly typed as samples are in DDS. However, zenoh values contain a field to specify which encoding is used as a MIME type. Any encoding may be used, but zenoh offers additional support for some encodings, which can be serialized and deserialized by the zenoh library itself. The encodings benefiting from this additional support include the following:

- **application/octet-stream:** The value is a stream of bytes.

- **text/plain; charset=utf-8:** The value is a UTF-8 string.

- **application/json:** The value is a JSON string.

- **application/properties:** The value is a string representing a list of key/value pairs separated by a semicolon (e.g., "id=002;squad=13...").

- **application/integer:** The value is an integer.

- **application/float:** The value is a float.

All zenoh values have a timestamp applied by the first router receiving it. A hybrid logical clock[2] generates this timestamp. It is a 64-bit value with a structure similar to NTP timestamps, although based on the UNIX epoch of January 1, 1970, UTC. It also contains the UUID of the router. Given the method used to generate them, zenoh timestamps are unique. In other words, each value existing in a zenoh system possesses a unique timestamp. The implication is that you can use the timestamps to order values at any location in the system without the need to leverage a consensus algorithm.

In router-less topologies, peer nodes can generate the timestamps themselves. In that case, they will use their own hybrid logical clock. To activate this behavior, all you need to do is to add the following setting to the JSON5 configuration file of the nodes:

```
add_timestamp:true
```

Primitives

The zenoh API is simple. There are only a few operations involved in data manipulations. The main operations are as follows:

- **get:** Executes a distributed query.

- **pull:** Pulls data for a subscription. This mechanism is useful to implement *pull subscribers*, that is, subscribers that are only awake or connected part of the time.

- **put:** Publishes a value for a key expression.

- **subscribe:** Declares a subscriber. You need to pass a reference to a callback triggered upon data reception.

There are also a few operations not focused on data itself. The most used of those operations are as follows:

- **close:** Closes a zenoh session.

- **open:** Opens a zenoh session.

---

[2] https://zenoh.io/docs/apis/rust

- **queryable:** Declares a distributed computation that will be triggered when a query is made. You need to pass a reference to a callback.

- **scout:** Searches for zenoh nodes on the local network. The type of nodes returned (clients, peers, routers) is specified through a bitmask.

Please remember that the actual names of the operations may vary, depending on the specific zenoh language binding you are using.

Deployment Units

When deploying zenoh infrastructure, you can mix and match different topologies according to your requirements. This level of flexibility is made possible by the specific features of zenoh node types.

There are three types of nodes in zenoh. Those types are as follows:

- **Client:** A node connected to one and only one peer or router. In zenoh, clients are often extremely constrained devices running zenoh-pico, a lightweight version of the protocol. Zenoh-pico can connect in client or peer-to-peer mode but cannot participate in mesh topologies.

- **Peer:** A node supporting peer-to-peer communication. Zenoh peers can route data on behalf of other peers. Moreover, zenoh clients can connect to a peer to communicate with the rest of the system.

- **Router:** A node able to route zenoh traffic between clients and peers, independently of the topology. Routers are especially useful in multisite deployments since they can connect over the public Internet without the need for complex networking setups. Zenoh routers also provide specific features through plugins, such as distributed storage instances. I will describe router plugins in the next section.

Figure 7-2 illustrates a few of the ways zenoh peer and router nodes can be deployed together. Mesh topologies, where peers connect directly and nonhierarchically to several other nodes, are supported. Cliques are simply a mesh where the nodes form a complete graph, in other words, a mesh where every node is connected directly with all others. Finally, peers can connect to one or several routers.

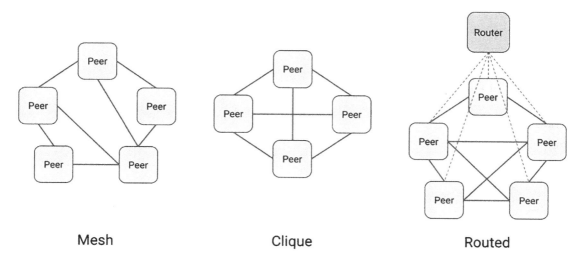

Mesh   Clique   Routed

Figure 7-2. *A few zenoh topologies*

Zenoh client nodes can connect to peers or routers. Zenoh peers can connect to other peers, routers, or both. This allows you to arrange nodes to meet your nonfunctional requirements. For example, deploying a mesh of peers in a specific location can remove the threat of a single point of failure in the infrastructure. Similarly, deploying redundant routers over distinct network connections can make multisite communications more resilient while enabling better system scaling. Figure 7-3 illustrates various potential topologies leveraging each of the three node types.

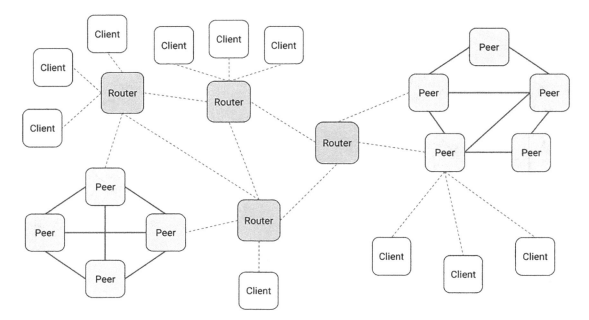

Figure 7-3. *Sample zenoh deployment*

One important detail about zenoh client and peer nodes is that their type does not constrain the operations they support. They can publish, subscribe, or both, depending on your requirements.

Device Discovery

Like DDS, zenoh implements powerful auto-discovery mechanisms. However, to keep the protocol efficient, its creators deliberately made efforts to minimize discovery traffic. Of course, the first step was optimizing the size of the discovery messages. They did not stop there, however. Zenoh only advertises resource interests, not specific publishers. Furthermore, such resource interests are generalized by the protocol itself; this means that if a device publishes data for values such as `/T800/sensors/lidar`, `/T800/sensors/camera`, and `/T800/position/latitude`, zenoh may decide to advertise the device under `/T800/**`. Finally, the reliability features of zenoh operate at the level of the nodes (or runtimes), not at the level of individual readers or writers like in DDS.

Overall, the generalization of resource interests in zenoh streamlines the discovery process and makes it possible for the protocol to support Internet-scale applications. All routing and matching leverage set-theoretic operations and set-coverage. This is useful, for example, to figure out the minimal set of storages that can answer a specific query. Zenoh performs resource generalization automatically, but you can use an API to provide hints to the infrastructure as needed.

Example: Robots at the Factory

Let's now illustrate what a real-world zenoh deployment would look like. Imagine your organization manages several factories on both sides of the Atlantic. Some are in Canada, and the others are in France. The factories are equipped with a video surveillance system and have autonomous robots working inside. A few router nodes provide the backbone of this system. Figure 7-4 provides a simplified view of the system.

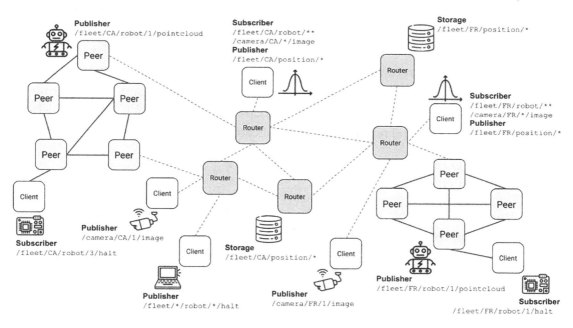

Figure 7-4. *A tale of two zenoh robot fleets*[3]

---

[3] This diagram contains icons from Flaticon.com by the following authors: Blak1ta, Eucalyp, Freepik, Good Ware, Lutfix, and IconBaandar.

The system designers decided to use /fleet as a prefix for all robot-related keys and /camera for all the video surveillance cameras. The second level of the key structure expresses the country of deployment, with the values CA for Canada and FR for France. In this system, robots are defined as peer nodes. Robots are not necessarily connected to every other one.

The robots constantly update their position by publishing to a sub-key named pointcloud. A level of the key also contains the robot's numerical ID, which is unique in the scope of the fleet. You can see two examples of the full key for that publication:

```
/fleet/CA/robot/1867/pointcloud
/fleet/FR/robot/1789/pointcloud
```

The robots contain two different computers: one is a generic single-board computer hosting applications, and the other a specialized board acting as the robot's control module. The control module is programmable and equipped with numerous I/O interfaces, enabling it to receive raw sensor data and control the robots' actuators. In this example, the pointcloud values are computed and published by an application running on the single-board computer. On the other hand, the control module is subscribing to commands sent by a central remote-control application on keys such as

```
/fleet/CA/robot/1867/halt
/fleet/FR/robot/1789/halt
```

For example, this remote-control application could prevent *Judgment Day* by publishing the appropriate value to the selector to stop all robots on the spot:

```
/fleet/*/robot/*/halt
```

Two clients, one in Canada and the other in France, subscribe to everything the robots and cameras publish. Specifically, they subscribe to the following selectors:

```
/fleet/<country code>/robot/**
/camera/<country code>/*/image
```

These analytics clients are connected to local routers; they aggregate position information for the robots and publish it under /fleet/<country code>/position/*. Two storages use the same key paths for their selectors. Consequently, a query made against /fleet/*/position/* would combine data from these two storages, which are attached to different routers than the analytics clients.

Plugins

Zenoh router nodes can provide features to the other nodes through plugins, which are libraries loaded by the router at startup or by administrators using the REST API at runtime. They share a runtime with the router, allowing them to call the zenoh Rust APIs with the same peer ID as the router process itself.

At the time of writing, the zenoh team maintained two plugins that provide core features in the infrastructure. The two plugins are as follows:

- **REST plugin:** This plugin implements the REST API for zenoh,[4] which you can use to manipulate the infrastructure or interact with the zenoh system to manipulate data. I will demonstrate how to use it later.

- **Storages plugin:** This plugin manages distributed storage instances. The plugin supports various storage *back ends* covering a variety of data management approaches.

Those two plugins ship in the standard zenoh distribution. The zenoh team maintains two additional plugins hosted in separate repositories. You can add them to your infrastructure as needed. The two plugins are as follows:

- **DDS plugin:** Scaling DDS traffic over a WAN or across multiple LANs is complex. This is due to the core design of DDSI-RTPS and its reliance on UDP/IP multicast packets. On the other hand, DDS recently saw tremendous adoption in the ROS 2 ecosystem. The zenoh DDS plugin[5] aims to encapsulate DDS traffic to streamline the integration of remote robot swarms. The plugin transparently maps DDS publications and subscriptions to zenoh resources. This blog post[6] explains in detail how to leverage the plugin.

- **Webserver plugin:** This plugin provides an HTTP server mapping URLs to zenoh keys. This makes it possible to serve files retrieved from distributed zenoh storage instances. A typical deployment leverages the filesystem storage back end to serve static files. However, the plugin will work with storages leveraging any of the supported storage back ends.

---

[4] https://zenoh.io/docs/apis/rest/
[5] https://github.com/eclipse-zenoh/zenoh-plugin-dds
[6] https://zenoh.io/blog/2021-04-28-ros2-integration/

Storages Plugin Deep Dive

Given the importance of distributed storage in zenoh, I will now cover in greater detail the storages plugin.

The core concepts implemented by the plugin are *storages* and *back ends*. Storage instances simply store key/values on a specific back end, and you need to assign them a selector at creation time. A back end is a storage technology supported by zenoh. Back ends abstract the details of the actual storage technology used. At the time of writing, the zenoh team was actively maintaining the following back ends:

- **filesystem:** The values are stored in a specific folder of the host's filesystem. Each value is a separate file, written in the same encoding used to transmit it. The key is reflected in the path for each file. See the back end's repository[7] for more details.

- **influxdb:** InfluxDB is a time-series database. Each storage instance corresponds to one database. It is possible to connect to an existing database or create a new one. The back end will create InfluxDB measurements for each of the keys. Values are recorded as InfluxDB points on these measurements. See the back end's repository[8] for more details.

- **memory:** Keys and values are stored inside an in-memory hashmap. The implementation of the back end is part of the storages plugin itself.

- **rocksdb:** RocksDB is an embeddable key/value store. The store resides in a file. Each storage corresponds to a specific database file. Zenoh keys and values are used directly, without mapping. See the back end's repository[9] for more details.

Zenoh back ends are packaged as libraries that are loaded by the storages plugin. There is no standard way to configure them, as the underlying technologies are too different. Back ends and storages can be created through a JSON5 configuration file loaded at startup by the router or the zenoh REST API.

---

[7] https://github.com/eclipse-zenoh/zenoh-backend-filesystem
[8] https://github.com/eclipse-zenoh/zenoh-backend-influxdb
[9] https://github.com/eclipse-zenoh/zenoh-backend-rocksdb

Each of the back ends provided has specific performance and persistence characteristics. The memory back end offers no long-term persistence but can be used as a fast cache in the infrastructure. The filesystem back end is easy to deploy but will not scale as a database would from a performance viewpoint. Using the InfluxDB back end makes sense if the system you build requires a historian database to support data analysis, as many SCADA use cases do. Finally, the RocksDB back end offers a scalable database engine that will deliver the best performance if you need to perform many random-access queries on the data.

The zenoh team also plans to introduce a back end supporting SQL databases in the future. Such a back end will provide better integration with typical IT infrastructure.

Reliability and Congestion Control

To understand how zenoh implements its reliability and congestion control features, you first need to know a bit more about zenoh's internals. The zenoh protocol is made of two distinct layers:

- **Session protocol:** The session protocol manages the one-to-one bidirectional sessions that nodes (runtimes) have with each other. Those sessions are the same whether clients, peers, or routers are involved. You can configure these sessions for best-effort or reliable delivery.

- **Routing protocol:** The routing protocol propagates interests and route data from producers to consumers. It leverages the session protocol.

Given how zenoh sessions work, the default reliability strategy is named *hop-to-hop* since delivery is guaranteed only between two specific nodes. This is sufficient in most cases since the infrastructure should be stable after its deployment. However, if nodes fail, then data loss is a possibility. Traffic will be rerouted, but there is no guarantee that all the in-flight data at the time of the outage will be delivered.

The hop-to-hop strategy is highly scalable but insufficient for some mission-critical applications. Consequently, zenoh offers two alternative strategies that are configured through APIs. The first one, *first router to last router*, relies on reliable channels between the first and last routers on each route. The second one, named *end-to-end*, is much less scalable and consumes more resources. This is because zenoh will create *reliable channels between every publisher and subscriber pair* in the infrastructure. Figure 7-5 compares the three reliability strategies available in zenoh.

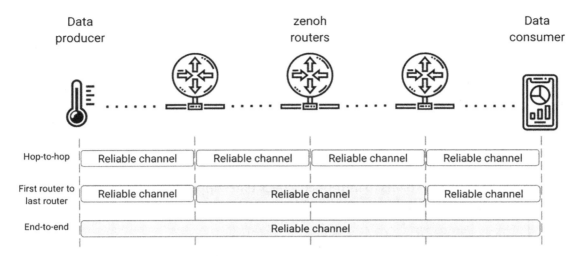

Figure 7-5. *Zenoh reliability strategies*

In addition to the reliability strategies it offers, zenoh separates control over message resending, memory usage (buffering), and message dropping. Subscribers control reliability by selecting a resending strategy. They declare if they need publishers to resend missing messages or not. On the other hand, publishers and routers individually decide how much memory they allocate to support reliability. This allows constrained devices to dedicate fewer resources, while routers will allocate more to alleviate congestion issues. Finally, publishers control what happens when congestion is detected by selecting a message dropping strategy. In zenoh, messages slated for reliable delivery are queued. When congestion happens, the memory dedicated to that queue is exhausted. Publishers need to specify if the infrastructure will drop the messages that cannot be queued or if the publication operation will be blocked. The message-dropping strategy is propagated from the publisher to all the relevant routers and applied to the entire routing path.

zenoh-flow

The zenoh protocol is a fantastic foundation for building all sorts of distributed systems, not just those related to IoT. A good proof of this is the zenoh-flow dataflow programming framework, which enables developers to deploy computations spanning from the Cloud to constrained devices.

Dataflow programming represents programs as a graph of connected nodes executing concurrently. When using dataflow programming languages, developers focus on the nodes and their interconnections, which express the application's logic. Such languages share some features of functional languages. The SystemVerilog and VHDL hardware description languages, commonly used to design processor cores and other electronic systems, are examples of dataflow programming languages.

In zenoh-flow, a dataflow is composed of three different types of nodes. Those three types of nodes are as follows:

- **operator:** This node implements a computation. It also defines input rules that determine what conditions will trigger the node and output rules that establish which output values will be passed to the following operators in the graph.

- **sinks:** This node consumes the data resulting from the dataflow.

- **sources:** This node produces the data that will be processed through the dataflow.

These nodes are loaded dynamically at runtime.

To work with zenoh-flow, you must specify the dataflow graph through a YAML file. You will use tags to express location affinity and requirements for the operators. The graphs can contain *loops*, supporting iterative processing and control mechanisms. They can also contain *deadlines* (timeout values), enabling you to leverage time-aware computations and implement proper fault tolerance. Graphs are also *composable*; in other words, they can include other graphs by reference. This makes them reusable and extensible. When you deploy the dataflow graph, zenoh-flow will automatically deal with the distribution of software components by linking remote operators through zenoh. At runtime, the sources, operators, and sinks leverage the zenoh protocol to communicate in a location transparent manner. They could be deployed in various ways to support specific requirements.

The zenoh team maintains a dedicated repository for zenoh-flow examples. You can browse it to learn how to define operators, sources, and sinks; it also contains examples of dataflow graphs in YAML. As of writing, zenoh-flow was still in alpha, which explains why I do not reproduce specific samples here.

zenoh Protocol Stack

A current trend among IoT protocols is to support a wider variety of network transports. MQTT can run on UDP with MQTT-SN, for example. Given this, it is not surprising that zenoh supports several transports out of the box since it is a recent piece of technology. Figure 7-6 shows the zenoh protocol stack.

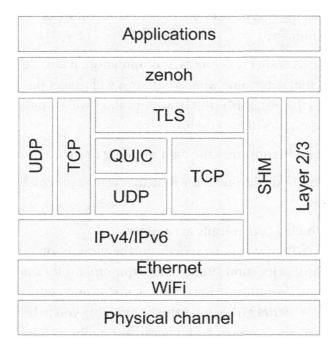

Figure 7-6. *The zenoh protocol stack*

What makes zenoh different is that it can run directly over an OSI level 2 data link without an intermediate transport. Of course, zenoh can also support TCP and UDP on top of IP. TLS is available for encrypted communications if you choose TCP. Moreover, you can also use TLS if you use the QUIC protocol, which relies on UDP. Finally, zenoh can also leverage shared memory and UNIX sockets transports.

Note   QUIC (pronounced "quick") is a general-purpose transport layer network protocol created by Jim Roskind at Google. QUIC leverages multiplexed UDP connections between two endpoints. QUIC is now a proposed standard at the IETF and is documented in RFC 9000.

The zenoh documentation contains detailed instructions about setting up TLS for use with TCP. The same configuration applies if you choose to use QUIC instead.

Coding with Eclipse zenoh

The core zenoh library is implemented in Rust. Consequently, the team publishes binaries for the main platforms supported by the language: Linux, macOS, and Windows. At the time of writing, the C and Python language bindings are supported on every platform where the target language and Rust are supported. This section will focus on Python code examples.

The official web resources for zenoh are as follows:

- **Website:** `https://zenoh.io`

- **Eclipse project page:** `https://projects.eclipse.org/projects/iot.zenoh`

- **Code repository:** `https://github.com/eclipse-zenoh/zenoh`

Installation

The first step to developing zenoh applications is to install Rust itself. Although there are binaries for the core libraries, you will need the Rust compiler to set up the language bindings.

Given that the zenoh team relies on the Cargo package manager, the easiest way to set up your Rust environment on Linux and macOS is to use the `rustup` script. You can do this by running this command:

```
curl https://sh.rustup.rs -sSf | sh
```

There is also a Windows version of `rustup` available; please look at Cargo's documentation for the latest instructions.

Core Library

The zenoh core library is available on macOS through the Homebrew package manager. To install the latest version, issue the following commands:

```
brew tap eclipse-zenoh/homebrew-zenoh
brew install zenoh
```

On Linux, the zenoh team maintains a private repository hosting packages in DEB format. You can use it to install zenoh on Debian and its numerous derivatives. Use the following commands to add the repository to your list of sources:

```
echo "deb [trusted=yes] https://download.eclipse.org/zenoh/debian-repo/ /"
| sudo tee -a /etc/apt/sources.list > /dev/null
sudo apt update
```

Then, you can use apt to install zenoh:

```
sudo apt install zenoh
```

You can download Windows binaries of the core library from the Eclipse Foundation's downloads website. The URL is

```
https://download.eclipse.org/zenoh/zenoh/latest/
```

The file's name will follow this pattern:

```
x86_64-pc-windows-msvc/zenoh-<version>-x86_64-pc-windows-msvc.zip
```

Building from the source is also possible.

C Binding

At the time of writing, the only way to install the C binding was to build it from the source. To get started, you need to clone the repository by executing this command:

```
git clone https://github.com/eclipse-zenoh/zenoh-c.git
```

Since Rust is already installed, we can build the code right away. On Linux and macOS, execute these commands:

```
cd zenoh-c
make
make install # on Linux prefix with sudo
```

You can also issue this command to build the examples:

```
make examples
```

Python Binding

There are two ways to install the Python binding: through `pip` or from the source. The binding has been tested with Python 3.6, 3.7, 3.8, and 3.9.

To install using `pip`, you will need at least version 19.3.1 of the package manager. If your environment fulfills this requirement, then all you need to do is to execute the following command:

```
pip install eclipse-zenoh
```

I recommend you leverage some sort of virtual environment.

To install from source, you need to clone the repository:

```
git clone https://github.com/eclipse-zenoh/zenoh-python.git
```

Once this is done, install the required dependencies by running these commands:

```
cd zenoh-python
pip install -r requirements-dev.txt
```

You can then issue this command to build and install the binding:

```
python setup.py develop
```

Testing Your Setup

The core zenoh library and the language bindings shipped by the zenoh team all include a comprehensive set of example programs. Those programs illustrate the most used functions in the API and are consistent from one implementation to another. I will now show you how to test your environment with the Python examples.

As DDS, zenoh supports peer-to-peer communication. So the easiest way to perform a test is to start a subscriber and publisher.

To start the subscriber, open a command line and navigate to the folder where you cloned the GIT repository of the Python binding. The examples can be found in the `examples` subfolder of the repository. Then, start the publisher by executing this command:

```
python z_sub.py
```

You should see this output:

```
Openning session...
Creating Subscriber on '/demo/example/**'...
Enter 'q' to quit...
```

Then, open another command prompt, go to the appropriate folder, and execute this command:

```
python z_pub.py
```

If all goes well, you should see output like this:

```
Openning session...
Declaring key expression '/demo/example/zenoh-python-pub'... => RId 1
Declaring publication on '1'...
Putting Data ('1': '[   0] Pub from Python!')...
Putting Data ('1': '[   1] Pub from Python!')...
Putting Data ('1': '[   2] Pub from Python!')...
```

And there should be new output on the side of the subscriber, showing that it received the publications:

```
>> [Subscriber] Received PUT ('/demo/example/zenoh-python-pub': '[   0]
Pub from Python!')
>> [Subscriber] Received PUT ('/demo/example/zenoh-python-pub': '[   1]
Pub from Python!')
>> [Subscriber] Received PUT ('/demo/example/zenoh-python-pub': '[   2]
Pub from Python!')
```

Zenoh's support for automatic discovery made it possible for the publisher and subscriber to communicate without the need for any configuration. Of course, this supposes the network does not block the traffic; you may need to adjust your firewalls if deploying nodes on different machines, depending on which of the supported transports you will use.

Testing the Zenoh Router

Most zenoh deployments will require at least a few routers. Consequently, it is useful to run router instances even on your development machine.

If you installed zenoh binaries from a repository, the router executable should already be on your path. If you compiled the zenoh core library from the source, the executables are in your repository clone's target/release subfolder. To start the router with an in-memory storage instance configured, simply execute this command:

```
./zenohd --cfg='plugins/storages/backends/memory/storages/demo/key_expr:"/demo/example/**"'
```

By default, the router will produce no console output. To verify it is working properly, you can call the REST API. Assuming the jq utility is installed, execute this command:

```
curl http://localhost:8000/@/router/local |jq
```

You should get output like this:

```
[
  {
    "key": "/@/router/AFF09D2102A34B96A4DF00DEF13A0F5F",
    "value": {
      "locators": [
        "tcp/172.19.19.252:7447"
      ],
      "pid": "AFF09D2102A34B96A4DF00DEF13A0F5F",
      "plugins": [
        {
          "name": "rest",
          "path": "/home/fdesbiens/zenoh/target/release/libzplugin_rest.so"
        },
        {
          "name": "storages",
          "path": "/home/fdesbiens/zenoh/target/release/libzplugin_
          storages.so"
        }
      ],
```

```
      "sessions": [],
      "version": "v0.6.0-dev-160-ga2f190f1 built with rustc 1.59.0
      (9d1b2106e 2022-02-23)"
    },
    "encoding": "application/json",
    "time": "None"
  }
]
```

You can see in the preceding JSON that the storages plugin has been loaded since I defined an in-memory storage instance at startup. By default, the router only loads the HTTP plugin. However, the memory back end is built into the storages plugin and does not require the loading of an external library to work.

You can use the following command to list the back ends present on a zenoh router:

```
curl 'http://localhost:8000/@/router/local/**/backends/*' |jq
```

With the configuration I used to start the router, you should get output resembling this:

```
[
  {
    "key": "/@/router/AFF09D2102A34B96A4DF00DEF13A0F5F/status/plugins/
    storages/backends/memory",
    "value": {},
    "encoding": "application/json",
    "time": "None"
  }
]
```

You can also use the REST API to list the storages defined on the router. This is the appropriate request:

```
curl 'http://localhost:8000/@/router/local/**/storages/*' |jq
```

The output should resemble this:

```
[
  {
    "key": "/@/router/AFF09D2102A34B96A4DF00DEF13A0F5F/status/plugins/
    storages/backends/memory/storages/demo",
    "value": {
      "key_expr": "/demo/example/**"
    },
    "encoding": "application/json",
    "time": "None"
  },
  {
    "key": "/@/router/AFF09D2102A34B96A4DF00DEF13A0F5F/status/plugins/
    storages/__path__",
    "value": "/home/fdesbiens/zenoh/target/release/libzplugin_storages.so",
    "encoding": "application/json",
    "time": "None"
  },
  {
    "key": "/@/router/AFF09D2102A34B96A4DF00DEF13A0F5F/status/plugins/
    storages/version",
    "value": "v0.6.0-dev-160-ga2f190f1",
    "encoding": "application/json",
    "time": "None"
  }
]
```

You will need to configure the router to leverage the other storage back ends. The zenoh repository contains a sample configuration file from which you can take inspiration. The filesystem, InfluxDB, and RocksDB back-end repositories also provide configuration information. It is also possible to use the REST API to define such storage instances if the router's runtime can load the appropriate libraries.

> **Note**   The zenoh router is also available as a container. Just run this command to deploy it: `docker run --init -p 7447:7447/tcp -p 8000:8000/tcp eclipse/zenoh`. Please note that since Docker does not support UDP multicast between a container and the host, you may have to tweak your configuration. The zenoh documentation provides the relevant details.

The zenoh REST API can be used to publish values and issue queries to the infrastructure. It is useful in real-world applications but also for testing and troubleshooting. For example, suppose that various nodes publish words of wisdom under the `/demo/wisdom/*` key and that a storage instance has been created to hold them. Send a PUT request to the appropriate endpoint to publish a value, as shown in the following. In this case, the value is a UTF-8 string published to the `/demo/wisdom/rome` key.

```
curl -X PUT -d 'Qui acceperint gladium, gladio peribunt.' http://
localhost:8000/demo/wisdom/rome
```

Conversely, the following HTTP GET request will return all available values:

```
curl http://localhost:8000/demo/wisdom/**
```

The following is a sample result:

```
[
{ "key": "/demo/wisdom/rome", "value": Qui acceperint gladium, gladio
peribunt., "encoding": "application/x-www-form-urlencoded", "time":
"2022-03-03T15:13:27.872392699Z/940FD467605F4D38934F3419A2C12E6F" }
]
```

Now that we have a working environment, time for some code! We will explore a few examples bundled with the Python language binding.

Scouting

The `scout` primitive operation in zenoh enables a node to determine which other nodes are available on the local network. The simplicity of the example program showcasing it is a testament to the power of the auto-discovery mechanisms that are part of the protocol. I reproduced that program:

```
import sys
import time
import argparse
import zenoh
from zenoh import WhatAmI

# initiate logging
zenoh.init_logger()

print("Scouting...")
hellos = zenoh.scout(WhatAmI.Peer | WhatAmI.Router, 1.0)

for hello in hellos:
    print(hello)
```

In this case, the program is looking only for peers and routers. The third parameter specifies the length of the scouting operation in seconds; adjusting this duration could help mitigate latency or catch sleeping nodes coming back online.

The following is the output of the program on my machine. At the time, I had only a router and a subscriber running.

```
Scouting...
Hello { pid: Some(6806944733A149B3B0A65991E9B43EC4), whatami: "peer",
locators: ["tcp/172.19.19.252:33293"] }
Hello { pid: Some(9863FA443A6749A89B4927D89EFF8716), whatami: "router",
locators: ["tcp/172.19.19.252:7447"] }
```

Subscribing

Creating a subscription on zenoh is nearly as simple as scouting. The subscription example program first defines a callback named listener that will be called every time a value is received. All it does is to print the key and value, as you can see at the top of the following listing:

```
def listener(sample):
    time = '(not specified)' if sample.source_info is None or sample.
    timestamp is None else datetime.fromtimestamp(
        sample.timestamp.time)
```

```
    print(">> [Subscriber] Received {} ('{}': '{}')"
          .format(sample.kind, sample.key_expr, sample.payload.
          decode("utf-8"), time))

# initiate logging
zenoh.init_logger()

print("Openning session...")
session = zenoh.open(conf)

print("Creating Subscriber on '{}'...".format(key))

sub = session.subscribe(key, listener, reliability=Reliability.Reliable,
mode=SubMode.Push)

print("Enter 'q' to quit...")
c = '\0'
while c != 'q':
    c = sys.stdin.read(1)
    if c == '':
        time.sleep(1)

sub.close()
session.close()
```

Establishing the subscription is a simple matter. All you need to do is open a zenoh session and call the `subscribe` method on the resulting object. In this case, the program specifies reliable delivery and a push subscription.

Publishing

Publishing with the zenoh Python API is as simple as calling the `put` method of a session instance. However, it is often preferable to declare the key expression and the publication before calling `put`. When you declare the key expression, zenoh will assign an integer ID to it and will use that ID in the publication messages. This saves bandwidth and improves throughput. Moreover, declaring the publication will ensure the node will not send messages if there are no matching subscribers currently online. This reduces bandwidth usage. The publication example program shows how to leverage key expression and publication declarations. I reproduced its core logic:

```
# initiate logging
zenoh.init_logger()

print("Openning session...")
session = zenoh.open(conf)

print("Declaring key expression '{}'...".format(key), end='')
rid = session.declare_expr(key)
print(" => RId {}".format(rid))

print("Declaring publication on '{}'...".format(rid))
session.declare_publication(rid)

for idx in itertools.count() if args.iter is None else range(args.iter):
    time.sleep(1)
    buf = "[{:4d}] {}".format(idx, value)
    print("Putting Data ('{}': '{}')...".format(rid, buf))
    session.put(rid, bytes(buf, encoding='utf8'))

session.undeclare_publication(rid)
session.undeclare_expr(rid)
session.close()
```

The program publishes a value every second in a loop by calling the put primitive method of the session object. The value is held in a string variable and written as UTF-8 bytes.

zenoh on Constrained Devices

While zenoh is lightweight as IoT protocols go, supporting its full feature set on extremely constrained devices is impossible. Consequently, the zenoh team created the zenoh-pico library, a lightweight C implementation of the zenoh client API. Zenoh-pico is well suited for deployment on devices running real-time operating systems.

Installation

You can download zenoh-pico binaries in DEB, RPM, and tgz format from the Eclipse Foundation's downloads website[10] if you run Linux. It is also possible to build the library from the source by cloning the repository. At the time of writing, zenoh-pico did not support Windows. All you need to do on supported platforms is execute make and make install (sudo make install on Linux) from the appropriate folder.

Zephyr Development Environment

To work with zenoh-pico on the Zephyr RTOS, the zenoh team recommends leveraging the PlatformIO platform. If you have it installed, you can execute the following commands to create the structure of your project:

```
mkdir -p /path/to/project_dir
$ cd /path/to/project_dir
$ platformio init -b reel_board
$ platformio run
```

You should, of course, substitute the identifier of your board of choice as the value passed to the -b switch. The zenoh team tested zenoh-pico with the following hardware: nRF52840 (such as the Adafruit Feather nRF52840), STM32 Nucleo-144 (Nucleo-F767ZI MCU), and Reel Board. Other boards will likely work, but you will need to tweak the configuration files provided by the zenoh team.

I will now show you how to proceed with the Reel Board. After installing PlatformIO and creating the project's structure, you will need to execute the following commands:

```
cp /path/to/zenoh-pico/docs/zephyr/reel_board/CMakelists.txt /path/to/
project_dir/zephyr/
$ cp /path/to/zenoh-pico/docs/zephyr/reel_board/prj.conf /path/to/project_
dir/zephyr/
$ ln -s /path/to/zenoh-pico /path/to/project_dir/lib/zenoh-pico
```

---

[10] https://download.eclipse.org/zenoh/zenoh-pico/

Once this is done, your PlatformIO project should have the following structure on disk:

```
project_dir
├── include
├── src
│   └── main.c
├── zephyr
│   ├── prj.conf
│   └── CMakeLists.txt
└── platformio.ini
```

You can then add your code directly to the main.c file or create additional files in the src folder as needed. To deploy the application once it is ready, just issue these two commands:

```
platformio run
platformio run -t upload
```

The easiest way to test such a program in your local environment is, of course, to have a zenoh router running.

Subscribing

The zenoh API is remarkably consistent across implementations. The following is a simplified zenoh-pico code to declare a subscription. I cut some declarations and removed part of the error handling code to keep the snippet short.

```
void data_handler(const zn_sample_t *sample, const void *arg)
{
    (void)(arg); // Unused argument

    printf(">> [Subscription listener] Received (%.*s, %.*s)\n",
            (int)sample->key.len, sample->key.val,
            (int)sample->value.len, sample->value.val);
}
...
int main(int argc, char **argv)
```

```
{
    zn_session_t *s = zn_open(config);
    znp_start_read_task(s);
    znp_start_lease_task(s);

    zn_subscriber_t *sub = zn_declare_subscriber(s, zn_rname(uri),
    zn_subinfo_default(), data_handler, NULL);

    ... // Some kind of loop here

    zn_undeclare_subscriber(sub);

    znp_stop_read_task(s);
    znp_stop_lease_task(s);
    zn_close(s);

    return 0;
}
```

As you can see, the general structure is similar to the Python sample I covered earlier. There is a callback declaration at the top of the file, which is referenced in the call to the zn_declare_subscriber function.

Publishing

Publishing in zenoh-pico follows the same general logic as with the other implementations of the zenoh API. You need to declare the publication with zn_declare_resource and then publish values by calling zn_write. I reproduced the most interesting part of the main function of the zenoh-pico publishing example:

```
zn_session_t *s = zn_open(config);
if (s == 0)
{
    printf("Unable to open session!\n");
    exit(-1);
}
```

```
// Start the receive and the session lease loop for zenoh-pico
znp_start_read_task(s);
znp_start_lease_task(s);

printf("Declaring Resource '%s'", uri);
unsigned long rid = zn_declare_resource(s, zn_rname(uri));
printf(" => RId %lu\n", rid);
zn_reskey_t reskey = zn_rid(rid);

char buf[256];
for (int idx = 0; idx < 5; ++idx)
{
    sleep(1);
    sprintf(buf, "[%4d] %s", idx, value);
    printf("Writing Data ('%lu': '%s')...\n", rid, buf);
    zn_write(s, reskey, (const uint8_t *)buf, strlen(buf));
}

znp_stop_read_task(s);
znp_stop_lease_task(s);
zn_close(s);

return 0;
```

As with the Python version, the program is looping and will publish one value
per second.

PART 2

Constrained Devices

## CHAPTER 8

The Hardware

Les bons outils font les bons ouvriers.

Good tools make good workers.

—Proverbe Français

The life cycle of IoT and edge computing solutions is counted in years, sometimes even in decades. Whether they are microcontrollers, gateways, or edge nodes, the hardware devices supporting them are held to higher standards than standard IT equipment. They will likely face harsh environmental conditions such as low or high temperatures, dust, electrical interference, humidity, radiation, and vibrations. Some form of ruggedization is a requirement most of the time. Moreover, such devices are often installed in hard-to-reach locations. Consequently, they need to be resilient and power efficient, especially if they operate over battery power.

If good tools make good workers, then good hardware makes good software. Given all of this, selecting the right hardware for a particular use case is critical. This chapter intends to provide you with guidelines to make that decision and survey the available choices in the market.

Selection Criteria

No two IoT deployments are the same. Consequently, attempting to come up with a definitive list of selection criteria for hardware would be foolish. In this section, I picked the most generic ones to give you a starting point in your evaluation process. Of course, you must attribute the proper weight to each criterion according to your priorities. I would recommend not to give overwhelming importance to the price, however. For your solution to be cost-effective over the long term, you will need a modicum of reliability and durability, and settling for bargain-basement hardware will counter that.

© Frédéric Desbiens 2023
F. Desbiens, *Building Enterprise IoT Solutions with Eclipse IoT Technologies*,
https://doi.org/10.1007/978-1-4842-8882-5_8

With that said, let's now look at the selection criteria.

Power Consumption

In most IoT solutions, constrained devices are deployed in the field to transmit data acquired by sensors and interact with the physical world through actuators. These devices draw power from a battery in many industries and use cases. However, a few alternatives emerged in the last few years. The maturation of Power over Ethernet (POE) infrastructure and standards make power delivery through network cables a possibility for solutions targeted at digital buildings and smart factories. At the other end of the scale, energy-harvesting technologies can provide very small amounts of power to low-energy electronics. For example, an energy-harvesting wall switch can send a short-range radio signal to turn on nearby lighting fixtures or devices.

As you can see, power consumption is an important concern for a wide array of IoT deployment scenarios. Of course, hardware is only part of the story when determining a specific component's efficiency. There are many other factors to consider, such as the type of sensors used, the sampling rate, networking technology and protocols used, and, finally, the software itself. However, hardware plays a prominent role in determining the power usage of a component. This includes the microcontroller (MCU) or processor used, the type of random-access memory, the type of persistent storage, and even the interfaces used to attach peripherals. Moreover, higher-powered components may produce a significant amount of heat, requiring cooling solutions that increase the overall power requirements. Striking the right balance between performance and power efficiency is essential.

Life Cycle

Given how long the life cycle of IoT and Edge solutions can be, having a stable supply chain of hardware components is critical. Rewriting software for a new hardware platform or new sensors brings little value to yourself or the end user. Fortunately, there are a few possible approaches to tackle this issue.

The first possible approach is to stick to proven suppliers with a track record of long-term product support. However, changes in product lines and strategy are common in the corporate world. Today's dependable suppliers can become unreliable over time. If your strategy relies on a specific component or board, securing supply in a legally binding agreement is essential.

An alternative would be to manufacture the components yourself. Traditionally, this was an avenue reserved for large organizations with specific manufacturing capabilities. Of course, microcontrollers and systems on a chip (SOCs) are complex integrated circuits that involve a significant amount of intellectual property. Therefore, several players in the market built very profitable businesses focused on IP licensing. Arm is, without a doubt, a prime example of this trend. The emergence of the RISC-V instruction set and open source processor IP based on it, such as the CORE-V family from the OpenHW Group, brings this strategy into the reach of organizations of all sizes.

Various software strategies can also help mitigate the issues related to changes in the hardware platform. Leveraging portable programming languages is one way to do it, although you should make sure any dependencies you leverage will be portable. Another possibility is to use hardware abstraction layers in your stack. The Eclipse MRAA and Eclipse UPM projects are good examples; MRAA provides a uniform API across multiple MCUs and boards, while UPM does the same for various sensors and actuators.

Over the long term, you will probably rely on a hybrid approach mixing trusted suppliers, open hardware, and hardware abstraction.

Use Case Requirements

IoT constrained devices and edge computing nodes are deployed outside the confines of the corporate data center. They are often positioned in hard-to-reach locations, where connectivity will be spotty and physical conditions harsh. A good example is the radiation-hardened processors deployed in deep space probes. While their specifications pale compared to cutting-edge embedded solutions deployed on Earth, they are uniquely well suited for their intended use.[1] In other words, the environment determines most of the nonfunctional requirements that the hardware needs to fulfill.

The availability of specific networking technologies and power sources at deployment sites will influence the hardware selection process. However, the functional requirements of the use case are equally important. Solutions for mission-critical, real-time applications will often require guaranteed latency for memory and network access. Memory protection and deterministic interrupt handling are also highly useful in applications where the device needs to process a great amount of sensor input while making decisions that impact human safety.

---

[1] See Jacek Krywko, "Space-grade CPUs: How do you send more computing power into space?" published on November 11, 2019, on arstechnica.com for an overview: https://arstechnica.com/science/2019/11/space-grade-cpus-how-do-you-send-more-computing-power-into-space/

A very important aspect to consider when selecting hardware platforms is the device connectivity requirements. You will need to connect the sensors and actuators you select to a constrained device or gateway. Consequently, the hardware platforms you choose must offer the appropriate type of I/O ports. The presence of external ports can prolong the useful life of equipment in some cases. For example, plugging a USB-based AI accelerator into a gateway is more cost-effective than completely replacing the device.

Of course, striking the right balance between fulfilling requirements and other considerations is often difficult. Mass market solutions such as the popular Raspberry Pi single-board computers are cost-effective and versatile but not particularly well suited for deployment in harsh environments. On the other hand, ruggedized components cost more and are often in more limited supply.

Security

Constrained devices, gateways, and edge nodes deployed in the field are vulnerable to a range of attacks that do not impact Cloud or traditional IT infrastructure. Their physical location exposes them to the risk of being physically compromised or even stolen. This can have serious consequences. A stolen device can be used to impersonate a legitimate one and reveal weaknesses in the server-side back end to attackers. The data stored on the device can also be used to expose trade secrets or even customer identities, with severe potential repercussions. You need a comprehensive security strategy for the devices you will deploy.

At the hardware level, such a strategy starts by leveraging a Trusted Platform Module (TPM) in devices that can host them. A TPM is a dedicated microcontroller designed to provide security-related functions. They can generate cryptographic keys and a nearly unforgeable hash key summarizing the hardware and software configuration of the device used in remote attestation. In addition, they can store cryptographic keys securely; those keys can be used in disk encryption or for authentication purposes, for example. TPMs, cryptographic keys, and digital certificates are building blocks you can use to implement *root of trust* in your devices. Root of trust is a chain of trust ensuring your devices boot only legitimate code.

Storage encryption, also called data encryption at rest, is a key feature for any device deployed in the wild. While software solutions address this need, it is also possible to offload this concern to hardware by using self-encrypting drives. Such a solution, of course, increases the cost and power consumption of the device. FIPS 140-2, Opal

2.0, and AES 256-bit hardware-based encryption are commonly available. FIPS drives are tamper resistant and are usually required for national defense and government applications transmitting controlled, unclassified information.

Physical security is also an important concern to prevent your devices from being stolen or compromised. Concretely, this means you should, at a minimum, be able to place a lock on the device enclosure. Securing devices in tamper-resistant cases is a good idea in most scenarios. Specialized cases can protect against a range of environmental factors. If this applies to your use case, you should look for NEMA or IP-rated cases. The NEMA rating system from the National Electrical Manufacturers Association rates enclosures on the protection they give from solid particulate ingress, liquid ingress, and corrosion. NEMA ratings are used mostly in the United States. Conversely, the IP rating system is used worldwide but evaluates only liquid ingress and solid ingress protection. NEMA and IP ratings can somehow be compared to each other, but direct comparisons are not possible since the two systems consider a different set of factors.

In most cases, it is good to remove or obstruct any ports, interfaces, or pins that are not actively in use. Attackers could use any of these to compromise a device.

Cost

Every one of the criteria I discussed will directly influence the cost of your devices. In return, the cost will undoubtedly drive many of your hardware decisions. Striking the right balance is not easy, hence the need for you to properly gather the functional and nonfunctional requirements for your projects. Moreover, at the time of writing, the market was still heavily impacted by the consequences of the COVID-19 pandemic. The so-called "chip shortage"[2] made many components hard to find, driving prices significantly upward. That temporary scarcity led many organizations to streamline their supply chain and, in some cases, ship products without specific features or use more expensive components.

If you are expecting to deploy devices at scale for yourself or your customers, you should explore the cost savings of bulk purchases. The implication is that you will reduce costs by standardizing on a smaller array of components. However, some could represent less optimal choices for specific applications.

---

[2] Will Knight, "Why the Chip Shortage Drags On and On … and On," Wired.com, November 12, 2021, www.wired.com/story/why-chip-shortage-drags-on/

If you are still in the prototyping or proof-of-concept phase, one way to alleviate the unavailability of hardware components is to use simulation tools. Such tools enable you to emulate specific boards and, in certain cases, sensors, actuators, and peripherals. In a blog post published in March 2022, Benjamin Cabé, my predecessor at the Eclipse Foundation, identified three interesting options in that space: Renode, Wokwi, and TinkerCAD.[3]

Microcontrollers

Microcontrollers (MCUs) are at the core of most constrained devices. They are small computers implemented on a single integrated circuit (IC), typically containing a processor core, random-access memory, timers, counters, and programmable input/output peripherals. MCUs are designed to support small control applications and deliver low power consumption at an affordable price. Usually, they offer flash or EEPROM storage sized from a few hundred kilobytes to a few megabytes. Random-access memory is small and typically ranges from 16 to 512 kilobytes. It is not unusual to see 8-bit, 16-bit, and 32-bit CPU cores in microcontrollers.

Note   You can see microcontrollers as a specialized, low-power system on a chip (SOC). The SOCs found in tablets, smartphones, and even Mac computers using Apple Silicon integrate higher-end features such as cellular modems, graphics processing units (GPUs), and even artificial intelligence accelerators.

Most of the microcontrollers currently available on the market rely on processor cores using some form of RISC architecture, with Arm and AVR being the most widespread options. However, RISC-V is starting to make inroads, and there are even open source core and MCU designs available such as the OpenHW CORE-V MCU.

Microcontrollers typically implement either the von Neumann architecture or the Harvard architecture. The term *von Neumann architecture* refers to a stored-program computer where instruction fetches and data operations cannot simultaneously occur since they involve a shared bus. The main alternative to von Neumann architecture is the

---

[3] https://blog.benjamin-cabe.com/2022/03/17/3-free-simulation-tools-to-work-around-the-global-chip-shortage

Harvard architecture. Harvard architecture computers are stored-program systems that possess a dedicated set of address and data buses for memory access and another set for instructions. The design of a von Neumann architecture machine is typically simpler than that of a Harvard one. Most current high-performance desktop-class and data center–class processors incorporate aspects of both architectures.

Arm: Cortex-M

The Arm architecture is unquestionably the dominant player in the microcontroller space today. The reason for this lies in Arm's strong focus on power consumption. The very first Arm processor, delivered in 1985, was designed to run at 1 W of power. The chip averaged under 100 mW typical power while delivering ten times the performance of Intel's 80286 or Motorola's 68020, which were popular processors at the time.[4]

Nowadays, Arm offers two specific design series that are a good fit for constrained devices: Cortex-M and Cortex-R. However, it would be best if you kept in mind that simply referencing the series and model is not enough to determine the full capabilities of a microcontroller. This is because Arm usually provides only the processor core intellectual property (IP), and the other capabilities will depend on the actual maker of the chip.

Note   ARM is a chip designer, not a chipmaker. It licenses its processor IP to others. Most of the processor core models it offers have several optional features. Therefore, knowing the specific Arm core in an MCU, for example, a Cortex-M4, is not enough. Please read specification sheets very attentively before selecting the hardware for your project.

---

[4] See https://en.wikichip.org/wiki/acorn/microarchitectures/arm1 for more details about this impressive chip.

Table 8-1 presents the Cortex-M cores and their high-level characteristics.

Table 8-1. *Arm Cortex-M cores and their characteristics*

Core	Arm Architecture	Computer Architecture	Pipeline
M0	Armv6-M	Von Neumann	3 stages
M0+	Armv6-M	Von Neumann	2 stages
M1	Armv6-M	Von Neumann	3 stages
M3	Armv7-M	Harvard	3 stages
M4	Armv7E-M	Harvard	3 stages
M7	Armv7E-M	Harvard	6 stages
M23	Armv8-M Baseline	Von Neumann	2 stages
M33	Armv8-M Mainline	Harvard	3 stages
M35P	Armv8-M Mainline	Harvard	3 stages
M55	Armv8.1-M	Harvard	4–5 stages

Covering in detail the myriad of options and variations available in the Cortex-M series would make this book thicker than it should be. I will now list the main features of each core.

- **M0:** This core is optimized for a small die size. Common in the most affordable MCUs.

- **M0+:** This core is a superset of the M0, enabling the use of the same toolchains. The pipeline is one stage shorter, resulting in lower power consumption. Some Cortex-M3 and Cortex-M4 features are available as options.

- **M1:** This core is optimized for integration into field-programmable gate arrays (FPGAs).

- **M3:** This core brings branch speculation to its three-stage pipeline, improving performance. It is used in microcontrollers and as a secondary core in larger chips.

- **M4:** This core adds DSP (digital signal processing) instructions to the Cortex-M3 and can be equipped with an optional floating-point unit (FPU). Cores that include the FPU are also known as Cortex-M4F.

- **M7:** This core features a six-stage superscalar pipeline with branch prediction. It can be equipped with a floating-point unit capable of single-precision and, optionally, double-precision operations. Cores that include the FPU are known under the Cortex-M7F designation. The M7 delivers almost twice the power efficiency of the M4.

- **M23:** This core resembles Cortex-M0+ and is optimized for a small die size. It adds integer divide instructions and TrustZone security features. It implements the newer Armv8-M architecture announced in 2015.

- **M33:** This core is a cross between the M4 and M23. It also offers the TrustZone instructions.

- **M35P:** This core improves the M33 by adding configurable parity and error correction code (ECC) memory features. It also sports a new instruction cache and implements tamper-resistance concepts.

- **M55:** This core is based on the newer Armv8.1-M architecture. It offers TrustZone instructions, DSP support, Coprocessor support, ECC, and many other features as options. It is a more modular and configurable core than any of its predecessors.

It is important to remember that Cortex-M cores do not necessarily implement all the instructions from the version of the Arm architecture they target. Moreover, the many optional features offered by Arm impact the instructions available. Generally, Cortex-M0, M0+, and M1 binaries will run unmodified on Cortex M3, M4, and M7 cores. Binaries targeting the Cortex-M3 will run as is on the M4, M7, M33, and M35P.

The 2021 edition of the Eclipse IoT and Edge Developer Survey found that 35% of respondents picked MCUs based on Cortex-M3 and Cortex-M4 cores. Chips featuring Cortex-M7 cores were picked by 30% of respondents, and 25% used microcontrollers spotting M0/M0+ cores. Respondents could pick up to three choices for that question.

AVR: The Heart of Arduino

Arm is in the licensing business. Since many architecture features are optional, you need to carefully read the documentation for the specific MCU you are procuring from a specific supplier. Things are much more straightforward with Microchip's AVR architecture since the company sells actual MCUs.

Note AVR has been known as Atmel AVR for a long time. Microchip acquired Atmel in 2016 and slowly discontinued the usage of the brand.

The AVR microcontroller architecture was launched in 1997. By 2003, Atmel had already shipped 500 million units. The introduction of the popular Arduino platform in 2005 significantly accelerated AVR's market adoption. AVR microcontrollers are modified Harvard architecture machines. Program and data are stored in different memory systems presented in different address spaces. However, the architecture offers special instructions that make it possible to read data from program memory.

AVR MCUs use 8-bit RISC processor cores. The instruction set has been expanded over time by introducing newer models. The AVR series can usually be clocked at up to 20 MHz, although specific models can reach 32 MHz. Higher clocks result in higher power consumption, of course. Recent models in the AVR family feature an on-chip oscillator, removing the requirement for external clocks or resonator circuitry.

From a storage perspective, AVR MCUs can offer up to 256 KB of internal, self-programmable instruction flash memory (384 KB on XMEGA) and an internal data EEPROM up to 4 KB. Random-access memory can reach up to 16 KB (32 KB on XMEGA). Specific models can also offer 64 KB of external little endian data space. Table 8-2 provides an overview of the characteristics of the main series in the AVR family.

Table 8-2. *Characteristics of the main series in the AVR family*

Series	Flash Size	Clock (MHz)	SRAM	EEPROM
tinyAVR	0.5–32 KB	1.6–20	64–3072 bytes	64–256 bytes
megaAVR	4–256 KB	1.6–20	256–16384 bytes	64–4096 bytes
AVR Dx	16–128 KB	20–24	4–16 KB	512 bytes
XMEGA	16–256 KB	32	1–32 KB	512–2048 bytes

Each of the series contains several products with different feature sets and characteristics. The following list highlights the differentiation factors for each series:

- **tinyAVR:** The ATtiny series is optimized for small package sizes and offers only limited peripherals.

- **megaAVR:** The ATmega series provides a wide range of I/O pins and more memory and storage.

- **AVR Dx:** The AVR Dx series contains multiple subseries focused on the conditioning of analog signals, functional safety, and human-computer interaction.

- **XMEGA:** The ATxmega series is the most powerful of the AVR family. The MCUs feature support for cryptography and offer analog-to-digital converters.

The most well-known consumer of AVR MCUs is the popular line of Arduino boards from the namesake company. Most current and past Arduinos feature MCUs from the ATmega series. For example, the Arduino Uno is equipped with an ATmega328P, and the Uno WiFi Rev 2 features an ATmega4809. The latter can run at up to 20 MHz and features 48 KB flash storage, 6 KB RAM, and 256 bytes of EEPROM.

RISC-V: A Serious Challenger

Introduced in 2010 by researchers at the University of California, Berkeley, the RISC-V architecture is an open standard. The documents for the instruction set architecture (ISA) are published under open source licenses, as were several core designs. The RISC-V Foundation was formed in 2015 to steward the evolution of the ISA; it relocated to Switzerland in 2019 to address concerns with US trade regulations and took the name RISC-V International as of March 2020.

RISC-V International permits unrestricted use of the ISA in hardware and software. This posture resulted in significant adoption. As of March 2022, 111 different cores, 31 SOC platforms, and 12 SOCs were listed on the organization's website. The actual market footprint of RISC-V is probably much higher since organizations will not necessarily publicize their use of the technology. It is worth mentioning that only members of RISC-V International have voting rights to approve changes and can use the trademarked compatibility logo.

One important characteristic of the RISC-V ISA is its modularity. The base integer instruction set is available in 32-bit and 64-bit formats (RV32I and RV64I). There are drafts for a 32-bit embedded (16 registers, RV32E) and even a 128-bit version (RV128I). All other types of instructions are documented in ISA standard extensions. Those extensions typically have a single letter for their name, although most recent ones have a name starting with Z followed by an alphabetical name. Widely used examples of standard extensions are integer multiplication and division (M), atomic instructions (A), single-precision floating point (F), double-precision floating point (D), and control and status register (Zicsr). Standard extensions are designed to work with all the standard base instruction sets and do not conflict with each other. Designers of RISC-V cores can also create their own nonstandard extensions to the instruction set. Industry collaboration initiatives regarding user-defined extensions have also started to emerge. A good example is the OpenHW CORE-V eXtension I/F, which aims to simplify the implementation of custom instructions and custom accelerators attached to or built into RISC-V cores. The RISC-V specification states that such extensions must have a name starting with "X" for user-level instructions and "SX" for supervisor-level ones.

Given this, surveying the whole RISC-V landscape would be difficult. This is because implementers can design their chips from scratch, adopt one of the many open source cores available, or even license open source or commercial cores and add proprietary features. On the one hand, this makes the ecosystem diversified, with cores, MCUs, and SOCs available for integration in constrained devices, gateways, edge nodes, and, increasingly, desktop-class and server-class hardware. On the other hand, this makes it harder to provide a high-level picture of the market, given the products' lack of common reference points. Nevertheless, there is a thriving ecosystem of RISC-V-based MCUs, and they are undoubtedly worth your consideration.

In 2021, the Eclipse IoT and Edge Developer Survey found that 9% of respondents used MCUs based on RISC-V technology. Even established players in the Arm market, such as Espressif, are announcing new products based on RISC-V.[5]

---

[5] See this press release from November 2020, for example: www.espressif.com/en/news/ESP32_C3

Choosing an MCU

With so many choices, selecting an MCU can be a daunting task. Ultimately, the requirements of your use case should guide your decisions. Here is a list of questions you should ask yourself:

- **Do you need to process more than a few events per second?** The primary advantage of 16-bit and 32-bit microcontrollers over 8-bit is their speed. If all you need to do is record the current temperature and humidity every ten seconds, then even an 8-bit microcontroller will do the trick. Of course, running an AI model integrating the real-time values of over a hundred sensors to perform predictive maintenance on a locomotive would require a higher-end MCU. Your use case will probably be between those two extremes.

- **Do you need to compute large numbers?** It is possible to compute large numbers on an 8-bit MCU. However, those calculations will be much faster on 16-bit and 32-bit hardware. If you need to perform image processing, video analytics, or even digital signal processing leveraging fast Fourier transforms, an 8-bit MCU running in the low MHz is completely out of the question.

- **Do you need advanced connectivity?** Driving advanced I/O interfaces, especially several of them simultaneously, requires more processing power than what an 8-bit MCU can provide. If you need Ethernet, Universal Serial Bus (USB), a Controller Area Network (CAN) bus, or multiple universal asynchronous receiver-transmitter devices (UARTs), then go for a 32-bit MCU.

- **What peripherals do you need?** Some MCUs offer the bare minimum, while others will add plenty of built-in peripherals and even sensors.

- **How critical is preserving battery life?** 32-bit microcontrollers consume more power than 8-bit ones. Of course, this can be somewhat mitigated by putting processor cores and other devices to sleep when not in use. However, 8-bit MCUs have a battery life advantage that explains their continued availability in the market.

Another dimension to that decision is whether you will use wireless connectivity or not. If you use a radio to connect to the network, it will likely constitute the most power-hungry component onboard, whether it involves BLE, LoRaWAN, LTE-M, or other technologies. Power usage will influence which connectivity option is the best if battery life is core to your use case.

- **What software does your target MCU support?** Not all MCUs benefit from the same level of software support. You need to consider software development kits (SDKs), libraries, and even real-time operating systems. Broader software support offers you more choice in building your solution.

Ultimately, it would be best if you always experimented in conditions as close to real-life one as possible before committing to a specific hardware platform for the long run.

Sensors and Actuators

Once you have selected your MCU, the next step is to find sensors and actuators that can interface with it and fulfill your requirements. Covering in detail the wide variety of sensors currently on the market would require a stand-alone book. I will rather focus on the various ways to interface sensors and actuators with your MCUs. But before that, let's look at the main characteristics you should watch when selecting sensors.

Sensor Characteristics

The quality of the measurements made by sensors is influenced by their design characteristics. Most of those influence the price and durability of the sensor. Here is a list of the main characteristics you should pay attention to:

- **Accuracy:** The accuracy reflects how well the value returned by the sensor conforms to the actual value. The accuracy is determined by several other factors, such as sensitivity, resolution, noise, systematic errors, and aging. In most cases, sensitivity and noise are more important than accuracy. You need to consider all possible sources of inaccuracy.

- **Measurement range:** This is the range of values for which the sensor can return accurate readings. Trying to measure values outside the measurement range will return incorrect signal values.

- **Reliability and precision:** A sensor should consistently return the same value for the same measurement. For example, when measuring the size of a piece of wood, the sensor should return the same value for consecutive measurements and measurements made after an arbitrary period. In other words, if the actual length of the piece of wood stays the same, the sensor should always return the same value. That said, environmental factors can insert variations in the readings. Specific sensors may also become less reliable and precise as they age. You can compensate for this in software, granted that you gathered enough information to do so through experimentation or direct experience with the sensor. Regular calibrations could also address the issue, although this is not possible for all sensors.

- **Resolution:** The sensor's resolution is the smallest difference that is capable of capturing. Suppose you wish to measure the distance to an object. The object is 2 meters away. If your sensor's resolution is 2 mm, the values it will return can be 1.998 m, 2.000 m, and 2.002 m.

- **Sampling rate:** The sampling rate, or sampling frequency, is how often measurements are taken. The sampling rate should be at least twice the frequency of the variations in the signal; this is called the Nyquist criterion.

- **Sensitivity and noise:** Those two factors are closely related. Sensitivity is the ratio between the output signal and the value measured. For example, when measuring ambient temperature in Celsius with a voltage output sensor, the sensitivity is expressed in V/c. On the other hand, noise is unwanted variations in the measured signal. It is often expressed as signal-to-noise ratio (SNR) for sensors. The noise level is affected by several external factors, such as electrical interference, temperature, and vibrations.

As you can see, you need to elicit and document your requirements properly to select the right sensors. Many issues in IoT-enabled industrial processes can be solved by using a different type of sensor or moving the existing sensor to a different location. Also, you should always deploy redundant sensors for critical steps in any process. The overall reliability of the solution will greatly outweigh the added costs.

Making the Connection

Like other computers, MCUs process information in digital form. In other words, they can directly interpret binary signals. Such signals can represent if a door is open or convey a button press. If the MCU operates at 3.3v, it will understand a 0v measurement as a binary 0 and a 3.3v measurement as a binary 1. Microcontrollers feature general-purpose input/output (GPIO) pins to send or receive digital signals. However, the real world is more nuanced than just zeros and ones. Therefore, many sensors and actuators work with analog values. In other words, they will send or process voltage values on a specific scale. For a 3.3v MCU, an analog sensor would return any value between 0v and 3.3v. Most MCUs are equipped with one or several analog-to-digital converters (ADCs) to make sense of such signals. ADCs can translate an analog voltage on a pin to a digital value. They can also convert voltage levels to a range supported by the MCU.

There are many different types of ADCs. The ADC on most Arduinos is a 10-bit ADC, meaning it can detect 1,024 (2^{10}) discrete analog levels. Some microcontrollers sport 8-bit ADCs ($2^8 = 256$ discrete levels), and some have 16-bit ($2^{16} = 65,536$ discrete levels) or even 24-bit ($2^{24} = 16,777,216$ discrete levels) ADCs. Another important characteristic of ADCs is their sample rate: the frequency expressed in Hertz (Hz) at which the ADC samples the input signal. It is critical to remember that sensors have their sampling rates as well. Consequently, you need to make sure the combination of sensors and ADCs you plan to use can deliver the sampling rate you need together.

Integrated circuit (IC) GPIOs are implemented in a variety of ways. The term designates pins dedicated to digital signals, not a uniform pin layout standard. Sometimes, expander ICs are used to interface GPIO pins to serial communication buses such as I²C and SMBus. Microcontrollers usually include GPIOs. Depending on the application, a microcontroller's GPIO pins may represent its primary interface to external circuitry, or they may be one type of I/O used among an array of options.

I²C, SPI, UART, and W1 are four highly relevant I/O interfaces commonly found on microcontrollers and even industrial computers. I will now summarize the characteristics of those four interfaces.

- **I²C (Inter-integrated Circuit):** A serial communications protocol common in MCU-based devices, particularly for interfacing with sensors, storage devices, and LCDs. It is a synchronous protocol since it uses a clock line. The I²C protocol is comparable to SPI but uses fewer wires. However, it can make working with multiple instances of sensors featuring fixed addresses difficult due to addressing collisions. I²C permits multiple main (master) nodes, which is not possible with SPI.

- **SPI (Serial Peripheral Interface):** Another synchronous serial communication protocol common in MCU-based devices. It is faster than both UART and I²C but requires more wires. Multiple subnodes can rely on the same SPI main pin, and devices can even be daisy-chained.

- **UART (Universal Asynchronous Receiver-Transmitter):** Otherwise known as a serial port, it is a commonly included peripheral on MCUs. Asynchronous serial ports provide a simple way for two devices to communicate without the need for a shared clock signal. UARTs require only two wires, although they frequently feature two additional ones (RTS and CTS) to implement hardware flow control, which prevents data loss. Whether hardware flow control is necessary depends on the wire protocol used.

- **1-Wire:** A protocol enabling low-speed (16.3 kbit/s) connections over a single wire. The wire also transits signaling and power. 1-Wire is similar in concept to I²C but delivers a longer range at the cost of reduced data rates. 1-Wire connections are typically one to many from the MCU to several peripherals.

You probably noticed that the four options I just discussed are serial and bidirectional protocols. This is no accident. Serial protocols reduce cabling requirements when deploying large numbers of devices, making them cost-effective. Table 8-3 compares the four protocols with each other.

Table 8-3. *Comparison of MCU I/O protocols*

	I²C	SPI	UART	1-Wire
Duplex	Half	Half – 3 wires Full – 4 wires	Full	Half
Server:Client(s)	*:*	1:*	1:1	1:*
Wires per device	2	3 or 4	1	1
Data rate	100 Kibit/s 400 Kibit/s 1 Mibit/s 3.4 Mibit/s	Depends on the devices (over 10 MiB/s is possible)	Up to 5 MiB/s	16 Kibit/s
Maximal distance	1m	0.2m	15m	Up to 350m (depends on the topology and number of devices)

Although I referred only to sensors in my explanations, the protocols I covered can also be used to control actuators. Another useful technique to leverage when working with them is pulse-width modulation (PWM). PWM is a technique used in robotics for driving motors and servos. Using internal counters, the MCU modulates the duty cycle of a square wave to control the quantity of power provided to a device. The average value of voltage (and current) provided to the device is controlled by turning the switch between supply and load on and off at a fast rate. Many MCUs ship with PWM controllers exposed to external pins as peripherals under firmware control.

Endpoints, Gateways, and Edge Nodes

Compared to MCUs, IoT endpoints (which often contain a single MCU), IoT gateways, and edge computing nodes are closer in capability to traditional servers. In fact, in some cases, edge computing nodes will be traditional servers deployed in a local data center at the edge of the fiber network of a connectivity provider. However, gateways and edge nodes are often found outside the confines of the data center. An IoT gateway could be used to integrate legacy programmable logic controllers (PLCs) over a CAN bus connection, for example. Edge nodes deployed in a wind firm could be used to aggregate data gathered from nearby turbines. For such deployment scenarios, relying on high-performance but low-power hardware makes much sense. Low-voltage processors can be cooled passively, and removing mechanical parts such as fans from the devices typically improves their mean time between failures (MTBF). Using low-voltage parts also has the added benefit of reducing heat output in the first place.

The 2021 edition of the Eclipse IoT and Edge Developer Survey discovered that 32% of respondents used processors based on the Arm v8 architecture in their IoT gateways and edge nodes. The AMD/Intel x86-64 architecture came in second place at 29%, and Arm v7 (32-bit) was the pick of 27% of respondents. Participants were allowed to select up to three choices for that question.

Given the dominance of the Cortex-A and x86-64 architectures, let's now look at the types of processors they have to offer.

Arm Cortex-A

Charting the Cortex-A ecosystem is no easy task, given the number of processor suppliers licensing the technology. Once again, rather than focusing on products, I will present the fundamental characteristics of the cores.

The Cortex-A family offers both 32-bit and 64-bit cores. Generally speaking, the 32-bit cores implement the Arm v7-A architecture except for the Cortex-A32 core, which implements Arm v8-A. Table 8-4 lists the main characteristics of 32-bit Cortex-A cores.

Table 8-4. *32-bit Arm Cortex-A processor characteristics*

Core	Arm Architecture	Characteristics
A5	Armv7-A	1 to 4 cores 4–64 KiB instruction and data caches (L1)
A7	Armv7-A	1 to 4 cores 8–64 KiB instruction and data caches (L1) 0–1 MiB L2 cache
A8	Armv7-A	Single core only 16–32 KiB instruction and data caches (L1) 0–1 MiB L2 cache
A9	Armv7-A	1 to 4 cores 16–64 KiB instruction/data caches (L1) 128 KiB–8 MiB L2 cache
A12	Armv7-A	1 to 4 cores 32–64 KiB instruction cache (L1) 32 KiB data cache (L1) 256 KiB–8 MiB L2 cache
A15	Armv7-A	1 to 4 cores per cluster 1 or 2 clusters per physical chip 32 KiB instruction/data caches per core (L1) Up to 4 MiB L2 cache per cluster
A17	Armv7-A	1 to 4 cores 32–64 KiB instruction cache (L1) 32 KiB data cache (L1) 256 KiB–8 MiB L2 cache
A32	Armv8-A	1 to 4 cores 8–64 KiB instruction/data caches (L1) 0–1 MiB L2 cache

64-bit Cortex-A cores implement the v8-A, v8.2-A, or v9-A architecture. They are commonly found not only in IT equipment but also in smartphones. The features of 64-bit Cortex-A cores are summarized in Table 8-5.

Table 8-5. *64-bit Arm Cortex-A processor characteristics*

Core	Arm Architecture	Characteristics
A34	Armv8-A	1 to 4 cores per cluster Multiple clusters per chip possible 8–64 KiB instruction/data caches (L1) 0 KiB–1 MiB L2 shared cache
A35	Armv8-A	1 to 4 cores per cluster Multiple clusters per chip possible 8–64 KiB instruction/data caches (L1) 128 KiB–1 MiB L2 shared cache (optional)
A53	Armv8-A	1 to 4 cores per cluster Multiple clusters per chip possible 8–64 KiB instruction/data caches (L1) 128 KiB–2 MiB L2 shared cache
A55	Armv8.2-A	1 to 8 cores per cluster Multiple clusters per chip possible 32–128 KiB instruction/data caches (L1) per core 64–256 KiB L2 cache 512 KiB–4 MiB L3 cache
A57	Armv8-A	1 to 4 cores per cluster 48 KiB instruction cache (L1) per core 32 KiB data cache (L1) per core 512 KiB–2 MiB L3 cache
A65	Armv8.2-A	1 to 8 cores per cluster Simultaneous multithreading (SMT) 16–64 KiB instruction/data caches (L1) per core 64–256 KiB L2 cache
A72	Armv8-A	1 to 4 cores per cluster 48 KiB instruction cache (L1) per core 32–64 KiB data cache (L1) per core 512 KiB–4 MiB L2 cache

(*continued*)

Table 8-5. (*continued*)

Core	Arm Architecture	Characteristics
A73	Armv8-A	1 to 4 cores per cluster 48 KiB instruction cache (L1) per core 32–64 KiB data cache (L1) per core 512 KiB–4 MiB L2 cache
A75	Armv8.2-A	1 to 4 cores per cluster 64 KiB instruction/data caches (L1) per core 128–512 KiB L2 cache 512 KiB–4 MiB L3 cache (optional)
A76	Armv8.2-A	1 to 4 cores per cluster 64 KiB instruction/data caches (L1) per core 128–512 KiB L2 cache 512 KiB–4 MiB L3 cache (optional)
A77	Armv8.2-A	1 to 4 cores per cluster 64 KiB instruction/data caches (L1) per core 256–512 KiB L2 cache 512 KiB–4 MiB L3 cache (optional)
A78	Armv8.2-A	1 to 4 cores per cluster 32–64 KiB instruction/data caches (L1) per core 256–512 KiB L2 cache 512 KiB–4 MiB L3 cache (optional)
A510	Armv9-A	1 to 8 cores per cluster Merged-core microarchitecture 32 or 64 KiB instruction/data caches (L1) per core 128 KB, 192 KB, 256 KB, 384 KB, or 512 KB L2 cache (optional) 256 KiB–8 MiB L3 cache (optional)
A710	Armv9-A	1 to 8 cores per cluster 32 or 64 KiB instruction/data caches (L1) per core 256 KB or 512 KB L2 cache 256 KiB–16 MiB L3 cache (optional)

Depending on the core selected, chipmakers can include up to four or eight cores per cluster and deploy several clusters on a single chip. Moreover, some of the Cortex-A cores can be deployed in a *big.LITTLE* configuration, where more powerful but power-hungry cores (big) are paired with power-efficient but slower cores (LITTLE). Processor designers can deploy the Cortex-A55 and A78 in such a configuration; the Cortex-A510 and A710 form another possible pair.

The Cortex-A65, A76, and A78 have AE variants (Automotive Enhanced) that add dual-core lockstep for safety applications. Such a configuration allows the cores to process the same tasks in parallel and compare results for discrepancies. This greatly enhances the functional safety of the solution.

Most chipmakers will ship their Arm products as systems on a chip, adding support for serial ATA (SATA), PCI Express (PCIe), USB, Ethernet, and lower-speed interfaces such as I²C, SPI, and GPIO, among others. Arm also offers to license its Mali series of graphics processing units (GPUs) that can accompany the processor cores on the chip package. Some players in the market license the processor cores from Arm and add their proprietary GPUs to the chips.

x86-64

Although there are only two main players in the x86-64 ecosystem, AMD and Intel, their product lines are plethoric and revised regularly. Rather than navigating ever-changing product lines, I will present the outlines of processors typically found in IoT gateways and edge nodes.

Both AMD and Intel have offerings targeted at the embedded market. Those are true systems on a chip bringing together processor cores, graphics processing units (GPU), and, in some cases, low-speed interfaces such as I²C, SMBus, SPI, UART, and others. Since they are close to desktop and server processors, they also offer interfaces associated with that class of machines, such as PCIe, SATA, and USB. Support for error correction code (ECC) memory is not uncommon. The higher-end and most expensive options integrate cores from the AMD EPYC and Intel Xeon families.

To control costs, many device makers will rely on processors hailing from the mobile product lines of AMD and Intel. In that case, they add circuitry to support the low-speed I/O interfaces needed separately. The advantage of such processors is that they offer lower power consumption and heat dissipation.

Of course, some edge computing workloads require the full power of server-class processors. What makes such devices edge devices then is not their use of embedded technologies but rather the physical location where they are deployed. However, it is typical to find servers suitable for deployment at the edge shipping in ruggedized cases or enclosures.

To provide you an idea of the types of x86-64 processors used in the market, here are the CPUs used in four specific devices that were shipping at the time of writing:

- **IoT gateway:** AMD Athlon Silver 3050e mobile processor (1.40 GHz, up to 2.80 GHz Max Boost, 2 Cores, 4 Threads, 4 MB Cache)

- **Industrial computer:** 8th Generation Intel Core i3-8145U mobile processor (2.10 GHz, up to 3.90 GHz with Turbo Boost, 2 Cores, 4 Threads, 4 MB Cache)

- **IoT gateway:** Intel Atom x5-E3940 embedded processor (1.60 GHz, up to 1.80 GHz boost, 4 Cores, 4 Threads, 2 MB cache)

- **Edge node (server):** Dual Xeon E5-2690v4 server processor (2.60 GHz, up to 3.50 GHz with Turbo Boost, 14 Cores, 28 Threads, 35 MB Cache)

Ultimately, you should pick the processor that fits your workload and target power consumption. Please note that server-class and embedded processors often benefit from longer market availability than alternatives, providing a stable hardware platform over the long term.

Open Source Hardware

This book focuses on open source technologies. It would not be complete without a discussion of successful open source hardware initiatives. There is a growing ecosystem of open source processor cores, microcontrollers, single-board computers, and sensors and actuators. I will provide examples for the first three categories in this section.

Because hardware strongly influences the design of IoT solutions, relying on open source components brings distinct advantages to device designers. IoT developers optimize their code for size and need to keep power consumption to a minimum – at least in the case of field-deployed devices. It is easier to reach the desired power profile when you control not only the software but also the hardware. Moreover, open source hardware makes it easier to have a stable hardware platform over the solution's life cycle; if one specific supplier discontinues the component, it is always possible to find another or contract a third party to provide it. Finally, open source hardware enables designers to extend and improve the components. Since IoT and edge computing solutions are often deployed in remote locations or harsh environments, the capacity to tweak the hardware to fit the use case is invaluable.

Processor Cores: OpenHW CORE-V Family

The Eclipse Foundation works closely with the OpenHW Group to build a comprehensive RISC-V ecosystem on a foundation of open source processor cores implementations. OpenHW Group is a Canadian not-for-profit, global organization where hardware and software designers collaborate to develop open source processor cores, related IP, tools, and software. All OpenHW ecosystem artifacts are developed in publicly accessible GitHub repositories. The core designs are defined in SystemVerilog and are licensed under the Solderpad Hardware License version 0.51. This license is based on the Apache License 2.0 and facilitates commercial adoption.

At the time of writing, all the currently available cores belong to the CORE-V family. The following is a description of the most important CORE-V cores. At a high level, the CVA family of CORE-V processors could power gateways and edge nodes from a use case perspective, while the CVE4 family targets microcontroller-class cores.

CVA6 Family

Originally known as the PULP Ariane core, the CORE-V CVA6 6-stage, single-issue, in-order core implements the RV32GC or RV64GC extensions with three privilege levels (M, S, U) to support a UNIX-like operating system fully.

CVE4 Family

The CVE4 family contains several cores. I am only describing the three most important ones.

- **CV32E40P:** Originally known as the PULP RI5CY core, the CORE-V CV32E40P is a 32-bit, 4-stage core that supports DSP operations and implements SIMD extensions among other features.

- **CV32E40X:** The CORE-V CV32E40X is a small and efficient, 32-bit, in-order RISC-V core with a 4-stage pipeline. The CV32E40X core is aimed at compute-intensive applications and offers a general-purpose extension interface that you can leverage to add external custom extensions to the core.

- **CV32E40S:** The CORE-V CV32E40S is a small and efficient, 32-bit, in-order RISC-V core with a 4-stage pipeline. The CV32E40S core is aimed at security applications and offers both machine mode and user mode, enhanced physical memory protection, and various antitampering features.

Microcontroller: CORE-V MCU

The OpenHW Group CORE-V MCU project aims to build a microcontroller leveraging the CV32E40P core. The MCU includes support for I²C, PWM, SPI, and UART, among other technologies. At the time of writing, the most convenient way to test drive it was to deploy the chip as a bitstream on Diligent's Nexys A7 or Genesys2 FPGA boards. However, OpenHW Group had plans to have development boards with an actual physical chip ready for distribution in 2023. Please monitor this page for more information regarding the CORE-V MCU DevKit project: `www.openhwgroup.org/core-v-devkits/`.

Note   Field Programmable Gate Arrays (FPGAs) are programmable semiconductor devices. In other words, they are chips that can be reconfigured to perform different logic functions. They are invaluable tools for chip designers, as they enable the testing of changes to a processor core or MCU design without the need to produce actual semiconductors.

Single-Board Computers

The builders of many popular single-board computers available currently made their designs available under open source licenses. I will now explore a few examples in detail.

Arduino

The popular Arduino 8-bit microcontrollers may spot proprietary AVR microcontrollers, but their overall design is open source. Detailed pinouts and schematics are available on Arduino's website and published under the Creative Commons Attribution-ShareAlike 4.0 International License.

The availability of such schematics gave birth to a robust ecosystem of compatible boards.

micro:bit

Next is the mighty BBC micro:bit. The micro:bit is an open source single-board computer leveraging Cortex-M MCUs. Launched in 2016, it has seen two different revisions up to now. The following is a list of the main characteristics for each version:

- **Version 1:** Nordic nRF51822 MCU, 16 MHz Arm Cortex-M0 core, 256 KB Flash storage, 16 KB RAM

- **Version 2:** Nordic nRF52833 MCU, 64 MHz Arm Cortex-M4 core, 512 KB Flash storage, 128 KB RAM

The micro:bit was initially designed by the British Broadcasting Corporation (BBC) and several partners. The program's goal is to encourage children to get actively involved with computers. Given this focus, the boards are equipped with LEDs, buttons, sensors (accelerometer, magnetometer, temperature, light), and Bluetooth connectivity. Version 2 adds a microphone, a speaker, and a touch sensor. The official boards feature a 25-pin edge connector with PWM outputs, GPIO pins, analog inputs, serial I/O, SPI, and I²C.

The Micro:bit Educational Foundation was created from the original BBC project in October 2016. The Foundation is a not-for-profit organization responsible for maximizing opportunities for students worldwide to learn with the micro:bit. Consequently, it shepherds the evolution of the micro:bit board and its related software stack. Although only the schematics and bill of materials are available for the official boards, the Foundation released an open source compatible reference design and extensive documentation. The reference design is licensed under the Solderpad Hardware License version 0.51. It offers the same components as the official version 1 board and is 100% software compatible.

Raspberry Pi Pico

While the Raspberry Pi single-board computers get much attention, the Pi Pico is a popular and well-documented microcontroller. It features an RP2040 chip designed by Raspberry Pi itself. The chip is equipped with two Arm Cortex-M0+ cores running at 133 MHz, 264 KB of SRAM, and supports up to 16 MB of off-chip Flash storage connected via a dedicated bus. The MCU offers 2 UARTs, 2 SPI controllers, 2 I^{2}C controllers, and 16 PWM channels. A low-resolution temperature sensor is also included.

The Raspberry Pi Pico board is equipped with 2 MB of Flash and exposes 26 GPIO pins, 3 of them being ADC capable. Although the RP2040 MCU is proprietary, the design files for the Pi Pico board are open source. There are several boards available from third parties that also use the chip.

CHAPTER 9

Connectivity

Nos vies sont faites de tout un réseau de voies inextricables, parmi lesquelles un instinct fragile nous guide, équilibre toujours précaire entre le cœur et la raison.

Our lives are made up of a whole network of inextricable paths, among which a fragile instinct guides us, an always precarious balance between heart and reason.

—Georges Dor, *Il neige, Amour…*

The network is at the core of any IoT or edge computing solution. Connectivity is what makes IoT different from core embedded computing, after all. Selecting and implementing the right type of network for a specific deployment is more complex than it looks. L Peter Deutsch and his colleagues at Sun Microsystems identified the potential pitfalls in the 1990s, known today as the fallacies of distributed computing. The list goes like this:

1. The network is reliable.

2. Latency is zero.

3. Bandwidth is infinite.

4. The network is secure.

5. Topology doesn't change.

6. There is one administrator.

7. Transport cost is zero.

8. The network is homogeneous.

© Frédéric Desbiens 2023
F. Desbiens, *Building Enterprise IoT Solutions with Eclipse IoT Technologies*, https://doi.org/10.1007/978-1-4842-8882-5_9

Since IoT and edge computing devices are located outside the confines of the Cloud or the corporate data center, you were probably already expecting the network to be unreliable and the bandwidth finite – unless you are just getting started. Moreover, you were probably not expecting latency to be zero since one of the aims of edge computing infrastructure is to reduce it. Nevertheless, the rest of the fallacies are still relevant. Security is still a huge challenge, network topologies change over time, there are several administrators, and the network will not be homogeneous.

In this chapter, I will explore a variety of networking technologies that you could use in a real-world deployment. After that, I will share network design considerations. Finally, we will look at some code for the Zephyr RTOS.

Connectivity Options

There are a great variety of options to connect your constrained devices to a network. Some of them directly support TCP/IP, which means that the devices can theoretically connect directly to the Internet. This is the case with cellular connectivity, Wi-Fi, and some narrowband options. Other technologies do not support TCP/IP, so you need to deploy a gateway to bridge the local device network with the Internet. Of course, some of the higher-level protocols I covered earlier can run over non-IP transports and still support Internet connectivity.

I will cover the most widespread alternatives in the market right now in this section. But how will you pick the right one for your project? Here is a list of a few factors to consider:

- **Hardware support:** Are there MCUs or gateways shipping with support for the technology you contemplate using? If not, are there radio modules on the market that can integrate with the hardware you plan to use?

- **Software support:** Does the RTOS or OS you plan to use support the networking technology and specific radio module you selected? Are the required libraries proprietary or open source? How mature is the networking stack?

- **Ecosystem:** How diverse is the supplier ecosystem for the networking technology you evaluate?

- **Range:** What effective range do you need to support your use case? Is it measured in meters, hundreds of meters, or kilometers?

- **Data rate:** Given the quantity of data you will transmit from and to your devices, what data rate do you need? Do you need to support mobility use cases involving pedestrians, cyclists, cars, or other vehicles?

- **Battery life:** Will your devices draw power from a battery? In the affirmative, what is your target battery life?

- **Interference:** Considering the physical location where the devices will be deployed, how much interference do you expect in the target frequency bands? Are there devices using the same technology or other technologies relying on the same frequency bands already deployed onsite?

- **Cost:** How much can you invest in network connectivity compared to other aspects of the solution? Given your constraints and requirements, what are the impacts of the network connectivity option you plan to use on the development time, maintenance efforts, and life cycle of the hardware and software?

Ultimately, you will probably need to compromise. The real world can be messy, especially when radio technologies are involved.

Bluetooth

Of all the technologies listed in this section, only Ethernet and Wi-Fi are as widespread as Bluetooth. Your phone, car, and laptop probably integrate this wireless technology already. Bluetooth is under the stewardship of the Bluetooth Special Interest Group. In Bluetooth, devices communicate using low-power UHF radio waves on a frequency band between 2.402 GHz and 2.483.5 GHz. This frequency band is globally unlicensed but regulated for industrial, scientific, and medical use.[1] In its most widely used mode, transmission power is capped at 2.5 milliwatts, resulting in a range of up to 10 meters (33 feet).

---

[1] Such bands are often called ISM (Industrial, Scientific, Medical) bands.

Note   The name *Bluetooth* refers to 10th-century Danish King Harald Bluetooth. It was proposed in 1997 by Jim Kardach of Intel, one of the founders of the Bluetooth SIG. Bluetooth was a placeholder name intended to be replaced before commercialization, but trademark issues got in the way. In any case, the name spread throughout the industry before it could be changed. The lesson here is always to select the code names of your projects carefully.

There are two types of Bluetooth technology: Bluetooth Low Energy (LE) and Bluetooth Basic Rate/Enhanced Data Rate (BR/EDR). Both types allow point-to-point connections between two devices. Both also operate in the same frequency band, but Bluetooth LE is the more popular option. It requires much less energy to operate and enables the creation of mesh networks. Bluetooth LE data rates can reach 1 Mbit/s (up to Bluetooth 4) or 2 Mbit/s (in Bluetooth 5), while BR/EDR can deliver up to 3 Mbit/s.

Bluetooth uses a radio technology called frequency-hopping spread spectrum. It is packet based and transmits each packet on one of 79 predefined channels. Each channel has a bandwidth of 1 MHz. Bluetooth Low Energy channels use 2 MHz spacing, accommodating 40 channels. Moreover, Bluetooth relies on a main/follower architecture. In BR/EDR, one main may communicate with up to seven followers to form a piconet: an *ad hoc* network in Bluetooth parlance. BLE allows theoretically millions of devices in a piconet, although going over a few hundred follower devices is impractical since all followers share the same radio time to communicate with the main. The main device in a piconet provides a clock used by the other devices as the base for packet exchange. Once a piconet is established, its members hop radio frequencies simultaneously to maintain the network and avoid interfering with other piconets that may be nearby.

Bluetooth BR/EDR requires devices to be paired. The pairing process establishes a trust relationship between the two devices, which then can exchange data over an encrypted connection. Bluetooth LE works differently. Pairing is possible but not mandatory. Bluetooth BLE devices can make themselves detectable by others by broadcasting advertising packets. This broadcast relies on three separate channels (frequencies) to reduce interference. The advertising device transmits an advertising packet on at least one of the three channels. It will repeat the transmission on a period called the *advertising interval*. A delay of up to ten milliseconds is added to each advertising interval to reduce the probability of multiple consecutive collisions. The actual duration of the delay is determined randomly.

Bluetooth is a popular choice for short-range connectivity in IoT solutions. It is available even in low-cost MCUs and supported by all serious real-time operating systems. Remember, however, that Bluetooth does not use TCP/IP, which reduces the range of protocols you can use over it. However, it is possible to establish IPv6 connections over Bluetooth by leveraging 6LoWPAN. 6LoWPAN is maintained by the IEFT and is documented in RFC 4944. It was updated by RFC 6282 (header compression) and RFC 6775 (neighbor discovery optimizations). 6LoWPAN over Bluetooth BLE is covered in RFC 7668.

On the downside, Bluetooth is known for well-documented privacy and security issues. Bluetooth security supports authentication and encryption. However, poor implementations are common, and the protocol is vulnerable to denial-of-service attacks, spamming, and backdooring. Security researchers, in the past, focused their efforts to identify Bluetooth vulnerabilities mostly on the pairing process. The increased scrutiny of other parts of the protocol led to the discovery of additional weaknesses. For example, in 2020, researchers from Purdue University uncovered issues in the Bluetooth BLE reconnection process, which unfolds when a device moves out of range and comes back online later. Many implementations make authentication optional during device reconnection, and the authentication process can be circumvented in specific circumstances.[2] You should thoroughly evaluate and test the security features of any Bluetooth devices you will deploy.

Cellular (LTE, 5G)

Wireless broadband communications are widely available in most populated regions of the globe. For this reason, cellular connectivity is a popular choice for many types of IoT and edge computing deployments. At the time of writing, LTE and 5G were the only two options that made sense since most wireless operators worldwide had started to wind down their 3G networks. 3G, LTE, and 5G are technologies managed by the 3rd Generation Partnership Project (3GPP). 5G is the successor technology to LTE.

---

[2] Catalin Cimpanu, "Billions of devices vulnerable to new 'BLESA' Bluetooth security flaw," zdnet.com, September 15, 2020. www.zdnet.com/article/billions-of-devices-vulnerable-to-new-blesa-bluetooth-security-flaw/

Long-Term Evolution (LTE) is a standard for wireless broadband based on the previous GSM/EDGE and UMTS/HSPA standards. LTE frequencies and bands vary from country to country; however, it is possible to procure devices and radios that support multiple bands. The bands will be found in a range from 450 MHz to 5 GHz. The LTE standard supports only packet switching over all-IP networks. Consequently, operators had to devise solutions to carry voice calls, such as the *Voice over Long-Term Evolution* (VoLTE) standard. The upside is that LTE and 5G networks support TCP/IP out of the box. Please do not confound LTE with LTE-M, a narrowband wireless technology I will cover later.

5G can be implemented in low-band, mid-band, or high-band millimeter-wave (mmWave) frequency ranges. Low-band 5G uses a range of 600–900 MHz, which overlaps with LTE. This results in download speeds of 30 to 250 Mbit/s, marginally better than LTE. The effective range of low-band 5G towers is the same as 4G towers. Mid-band 5G, on the other hand, relies on microwaves in the 1.7 to 4.7 GHz range. This translates to speeds between 100 and 900 Mbit/s. The mid-band 5G cell towers' range can reach several kilometers in radius. Finally, high-band 5G operates in frequencies currently ranging from 24 to 47 GHz, although higher frequencies could be used in the future. mmWave 5G can theoretically deliver speeds over a gigabit per second (Gbit/s), comparable to the best wired Internet connections. However, waves in such high frequencies are easily blocked by the construction materials used in walls or windows. Consequently, the effective range of the towers is much smaller, driving up the costs of deployments. mmWave 5G is thus scarce outside densely populated urban areas or venues where large quantities of people congregate, such as sports stadiums.

LTE and 5G connectivity is often found in gateways and edge nodes, inside which the proper radio modules are often integrated through a USB or PCIe connection. As with smartphones, the radio module requires a SIM (Subscriber Identity Module or Subscriber Identification Module) card to work properly. Such cards are integrated circuits designed to securely store the international mobile subscriber identity (IMSI) number and its corresponding key. Both are used to identify and authenticate subscribers on wireless broadband networks. In recent years, eSIMs (embedded SIMs) have gained in popularity. eSIMs are surface mounted permanently inside a device and are programmable; it is even possible to provision them remotely. This improves the reliability and security of the devices.

Many gateways offer support for two SIMs or more, enabling the use of an alternate network as a fallback in the case of issues with the primary network. LTE and 5G connectivity can also be used as an out-of-band connectivity solution to manage devices; it can be cost-effective in such scenarios since it is usually invoiced according to bandwidth consumption.

Ethernet

Ethernet has been the dominant standard used in wired local area networks (LAN), metropolitan area networks (MAN), and wide area networks (WAN) since the 1980s. By the end of that decade, it had supplanted competing alternatives such as ARCNET and Token Ring. The IEEE Standards Association maintains the various versions of the Ethernet standard.

The original 10BASE5 Ethernet used a coaxial cable attached to every machine on the network (shared media), while newer variants use twisted pair and fiber-optic links in conjunction with switches. The most common forms used currently are 10BASE-T, 100BASE-TX, and 1000BASE-T. They offer speeds of 10 Mbit/s, 100 Mbit/s, and 1 Gbit/s. A higher-performance version of the standard, 10 Gigabit Ethernet (10GE, 10GbE, or 10 GigE), also exists but has seen more limited adoption than previous versions since it is still more expensive on a per-port basis. Given this, the 2.5GBASE-T (2.5 Gbit/s) and 5GBASE-T (5 Gbit/s) standards emerged to deliver higher throughputs than 1000BASE-T without the cost associated with 10GE.

The maximum length allowed for Ethernet cables depends on the cabling technology used. 100BASE-TX and 1000BASE-T rely on twisted pair cables. They both specify a maximum cable length of 100 meters (330 ft) and required Category 5 cable at their introduction. Since then, IEEE deprecated Category 5 cable; new installations use Category 5e. Fiber-optic variants typically use small form-factor pluggable (SFP) hot-pluggable network interface modules. Depending on the type of cable, connector, and version of the standard selected, the maximum cable length goes from 550 m (about 1800 feet) to 160 km (about 100 miles).

Ethernet is not limited to data transmission. Power over Ethernet (PoE) describes several standards or ad hoc systems that transmit electric power and data on twisted-pair Ethernet cabling. There are several widely used techniques for transmitting power over Ethernet. Three of them have been standardized as IEEE 802.3 since 2003. The original IEEE 802.3af-2003 PoE standard delivers up to 15.4 W of DC power (minimum

44 V DC and 350 mA) to each port. The device will only have access to 12.9 W due to power dissipation. The updated IEEE 802.3at-2009 PoE standard (often called PoE+ or PoE plus) delivers up to 25.5 W of power. This standard prohibits using all four pairs of wires in the cable for power. IEEE incorporated 802.3af-2003 and 802.3at-2009 into the IEEE 802.3-2012 publication. Finally, the IEEE 802.3bt-2018 standard (also known as PoE++ or 4PPoE) pushes further the capabilities of 802.3at. This standard introduces two additional power types: up to 51 W delivered power (Type 3) and up to 71.3 W delivered power (Type 4). Each pair of wires needs to handle 600 mA (Type 3) or 960 mA (Type 4). All IEEE PoE variants support Gigabit Ethernet (1000BASE-T), and 802.3bt can work over 2.5GBASE-T, 5GBASE-T, and 10GBASE-T.

Nowadays, most desktop computers ship with support for Gigabit Ethernet, as do nearly every IoT gateway and edge node. Many MCUs can also ship with Ethernet connectivity. For example, the Arduino Ethernet combines an ATmega328 with a W5100 Embedded Ethernet Controller from WIZnet, which delivers support for 10/100 megabits twisted-pair Ethernet connectivity. It is possible to add an IEEE 802.3af compliant PoE module to power the board. Raspberry PI also ships an 802.3af PoE module that can be used to source power for the Raspberry Pi 4 Model B and the Raspberry Pi 3 Model B+ boards. Of course, any network switch you will use with PoE-enabled devices will need to support the correct version of IEEE 802.3. Alternatively, it is possible to use standard Ethernet switches and deploy PoE injectors, adding power to the Ethernet signal delivered by a non-PoE device.

Ethernet is a proven solution to deliver wired connectivity to many IoT and edge computing devices. It is especially well suited to equipment deployed in fixed locations indoors. PoE is very common in digital buildings and connected factories, among other settings.

Narrowband

A low-power wide-area network (LPWAN) is a wireless network aimed at battery-operated constrained devices and sensors. It is designed to favor range over throughput to optimize the battery life of the devices. Most LPWANs exhibit data rates between 0.3 kbit/s and 170 kbit/s per channel. It is possible to set up your own LPWANs or use services and infrastructure deployed by third parties. The advantage of the second approach is that you will not have to deploy and maintain gateways in the field.

There are several competing standards in this market. I focused on four specific options: DASH7, LoRaWAN, LTE-M, and NB-IoT. All narrowband options currently in the market offer a long range, low power consumption, and cost-effectiveness. The operating range of LPWANs can reach over 10 km (6 miles) in rural settings; in densely populated urban landscapes, it will still reach a few kilometers. As for power consumption, battery life of over a decade is not unheard of. Moreover, LPWANs rely on topologies that keep infrastructure costs in check. They also use lightweight protocols that lower hardware complexity, thus reducing the cost of the devices. Some popular options rely on proprietary technologies that tend to result in more costly transceivers.

DASH7

The DASH7 Alliance Protocol (D7A) is an open source wireless narrowband protocol that operates in the 433 MHz, 868 MHz, and 915 MHz unlicensed frequency bands. It is placed under the stewardship of the DASH7 Alliance, a nonprofit mutual benefit corporation formed to foster the further development of the DASH7 protocol specification. You can download the specification for free from the DASH7 website.[3] The Eclipse Foundation is proud to be a member of the DASH7 Alliance.

D7A is derived from ISO/IEC 18000-7, a standard defining an interface for active radio-frequency identification (RFID) in the unlicensed 433 MHz ISM band. D7A no longer complies with ISO/IEC 18000-7. The protocol specification is free to use without any patent or licensing requirements. The range for D7A is short. It topples at around 500 m (1640 feet) for industrial applications, although device makers advertise read ranges of over a kilometer (0.62 miles) in line-of-sight situations. The protocol offers a typical latency of one second while consuming only 30 uA on average. There are three tiers of data rates, depending on the radio modulation selected. DASH7 low rate delivers an equivalent bitrate of 9.6 kbit/s, DASH7 normal rate reaches 56 kbit/s, and DASH7 high rate achieves 167 kbit/s.

The Sub-IoT project on GitHub houses an open source implementation of D7A. It is available under the Apache License v2.0.

---

[3] www.dash7-alliance.org/product/dash7-alliance-protocol-specification-v1-2/

LoRaWAN

LoRaWAN involves two distinct technologies: LoRa and LoRaWAN.

LoRa (derived from "Long Range") is a proprietary radio modulation technique. It was developed and patented by Cycleo, a company based in Grenoble (France), which Semtech later acquired. LoRa chipsets are provided exclusively by Semtech. LoRa uses the following unlicensed sub-gigahertz radio frequency bands:

- Asia: AS923 (915–928 MHz)

- Europe: EU868 (863–870/873 MHz)

- India: IN865 (865–867 MHz)

- North America: US915 (902–928 MHz)

- South America: AU915/AS923-1 (915–928 MHz)

- Worldwide: 2.4 GHz

LoRa can deliver data rates between 0.3 kbit/s and 27 kbit/s. The effective range can be up to 5 km (3.1 miles) in urban areas and reach over 15 km (9.3 miles) in rural areas, where unimpeded line of sight is easier to achieve.

LoRaWAN, on the other hand, defines the software communication protocol and system architecture. It is developed and maintained by the LoRa Alliance, a nonprofit established in 2015. LoRaWAN manages the frequencies, data rate, and power for all devices. The devices rely on the reporting by exception pattern: they transmit data asynchronously when needed. Those transmissions are forwarded to a centralized server by gateways deployed in the field. LoRaWAN connections are bidirectional, and the protocol supports multicast addressing groups to make the mass distribution of messages to devices more efficient. This is especially useful for distributing over-the-air software updates.

There are several open source LoRaWAN software stacks available. One of them is provided by the ChirpStack project, which ships a gateway bridge, network server, and application server, among other components. All components are available under the MIT license.

The long range and wide industry support for LoRaWAN make it an attractive technology for large deployments of wireless sensors with limited data rate requirements. Smart cities and smart agriculture are two good potential use cases. However, the proprietary nature of the radio modulation is a source of concern.

LTE-M and NB-IoT

LTE-M (or LTE-MTC, for "Machine Type Communication") and NB-IoT (Narrowband Internet of Things) are two LPWAN radio technology standards developed by the 3GPP to address machine-to-machine and IoT use cases.

LTE-M relies on existing LTE networks but is not necessarily available everywhere LTE connectivity is. Compared to an LTE modem, an LTE-M model will be less expensive and offer a battery life of several years. LTE-M networks also offer greater coverage and better indoor penetration than LTE ones. There are two categories of LTE-M: Cat-M1 and Cat-M2. Cat-M1 offers data rates of 1 Mbit/s for both uplink and downlink; Cat-M2, on the other hand, offers 4 Mbit/s of downlink and 7 Mbit/s of uplink. Latency averages between 50 and 100 milliseconds. Since it relies on cellular towers, LTE-M supports mobility use cases and supports VoLTE voice services.

Like LTE-M, NB-IoT uses a subset of the LTE standard but limits the bandwidth to a single narrowband of 200 kHz. The technology focuses on cost-effective coverage of indoor spaces where achieving a high density of connections is possible. Battery life of more than ten years can be supported for a wide range of use cases. There are two categories of NB-IoT. Cat-NB1, the first version introduced, delivers downlink data rates as fast as 26 kbit/s and uplink data rates up to 62 kbit/s. Latency ranges from 1.6 to 10 seconds. Cat-NB2, the newest version, pushes the downlink data rate to 127 kbit/s and the uplink data rate to 150 kbit/s. NB-IoT also supports Radio Resource Control (RRC) connection reestablishment, allowing devices to transfer their connection from one cell to another cell. This improves the technology's suitability for mobility use cases, such as pedestrian, cyclist, and vehicle monitoring.

Compared to NB-IoT, LTE-M provides better data rates and lower latency. Consequently, it is a better choice for real-time use cases. Its support for voice communications enhances its flexibility. Moreover, since LTE-M is an extension of LTE, it supports roaming, whereas NB-IoT does not. NB-IoT is the superior technology for any use case involving connecting to devices located in warehouses, underground locations, and commercial buildings. It is also a great choice if you only need to transmit sensor data at a low data rate.

Both LTE-M and NB-IoT involve SIM cards. However, the availability of LTE or 5G connectivity at a specific location does not guarantee that LTE-M and NB-IoT will be available. There are countries where one option is available, and the other is not. The GSM Association (GSMA) website features a useful map illustrating the worldwide commercial availability of LTE-M and NB-IoT services. You can find it here: `www.gsma.com/iot/deployment-map/`

Wi-Fi

Wi-Fi is a group of wireless network protocols based on the IEEE 802.11 standards. It is designed to integrate seamlessly with Ethernet. The evolution of Wi-Fi is driven by the Wi-Fi Alliance, a nonprofit organization. At the time of writing, Wi-Fi had achieved remarkable market adoption, with over 18 billion devices in use and 4.4 billion devices expected to ship in 2022.[4]

The first version of the 802.11 protocol was introduced in 1997. Since then, several successive versions of the standard have been introduced. Each specific version of the standard determines the radio bands used, the maximum ranges, and the maximum speeds that could be achieved. Wi-Fi most commonly uses the 2.4 GHz and 5 GHz radio bands; these bands are subdivided into several channels. Table 9-1 summarizes the main features of each version of Wi-Fi.

Table 9-1. *Versions of the Wi-Fi standard*

Generation	IEEE Standard	Maximum Data Rate (Mbit/s)	Ratification	Frequency (GHz)
Wi-Fi 0	802.11	2	1997	2.4
Wi-Fi 1	802.11b	11	1999	2.4
Wi-Fi 2	802.11a	54	1999	5
Wi-Fi 3	802.11g	54	2003	2.4
Wi-Fi 4	802.11n	600	2008	2.4 and 5
Wi-Fi 5	802.11ac	6933	2014	5
Wi-Fi 6	802.11ax	9608	2019	2.4 and 5
Wi-Fi 6E			2020	2.4, 5, and 6
Wi-Fi 7	802.11be	40000	2024 (expected)	2.4, 5, and 6

Please note that the Wi-Fi Alliance only started to use generational numbering in 2018. Older materials refer to the name of the relevant IEEE standard instead. Moreover, the Wi-Fi Alliance never referred publicly to generation numbers 0 to 3; those generation numbers have been inferred from the fact that the Alliance refers to 802.11n as Wi-Fi 4.

---

[4] www.wi-fi.org/news-events/newsroom/wi-fi-alliance-2022-wi-fi-trends

The maximum operational range of a Wi-Fi deployment depends on several factors. These include the frequency band, the radio power output, the sensitivity of the receivers used, the modulation techniques, and the gain and type of antenna used. The data rate is usually lower at longer distances and with greater signal absorption from walls and other obstacles. Lower radio frequencies will result in better range but lower throughput. Building materials absorb 5 GHz signals to a greater degree than 2.4 GHz ones. However, higher frequencies offer a greater number of channels, helping to reduce interference in environments where multiple networks compete for the spectrum. It is possible to increase the range of a Wi-Fi network by deploying multiple access points or by using range extenders and signal amplifiers. Recent Wi-Fi generations enable the deployment of mesh networks, improving coverage inside buildings, for example.

Many single-board computers and MCUs offer support for Wi-Fi. However, while it offers a greater effective range than Bluetooth, it also features much greater power consumption. To address this, the IEEE introduced the 802.11ah standard, also called Wi-Fi HaLow by the Wi-Fi Alliance. 802.11ah uses the sub-gigahertz ISM spectrum. The bands vary by region and are located between 755 and 928 MHz. The 802.11ah standard defines several modulation schemes and coding rates; there is a wide array of possible data rates from 150 kbit/s to 86.7 Mbit/s. The typical range is over 1 km (0.6 miles), and the protocol integrates multiple features to preserve battery life.

Wi-Fi is certainly a viable option for IoT and edge computing deployments, especially if your devices and sensors are not battery-powered. However, its limitations in range mean there are better options for outdoor deployment over large surface areas.

Z-Wave

Z-Wave is a wireless communications protocol used primarily in smart home deployments. It resembles Bluetooth BLE since it is essentially a mesh network using low-energy transmissions. The Z-Wave Alliance was established in 2005 to manage the evolution of the standard. However, Z-Wave transceiver chips are available exclusively from Silicon Labs, the current owner of the technology, and its licensees.

Z-Wave is designed to transmit small data packets with low latency at data rates up to 100 kbit/s. The maximum distance between two nodes on the network is between 30 and 40 meters (98 to 131 feet). Packets can hop from one node to another up to four times, increasing the potential coverage area. Z-Wave uses frequency-shift keying (FSK) modulation with a Manchester encoding. Like DASH7 and 802.11ah, it uses the ISM band. Z-Wave transmits at 868.42 MHz in Europe and 908.42 MHz in North America; it uses other frequencies elsewhere as defined by local regulations.

At a minimum, a Z-Wave network must contain a primary controller and at least one device. Controllers typically take the form of wall switches and handheld controllers, although software applications can also play that role. Networks can contain up to 232 devices, and it is possible to bridge several networks together. Devices need to be paired (or added) to the network before they can be controlled through the protocol. Given the power-saving features defined in the standard, the expected battery life for most devices is several years.

Z-Wave offers a security layer that delivers message integrity and confidentiality. It also implements mechanisms to ensure data freshness, meaning that the infrastructure will only process messages sent recently. Z-Wave relies on an AES symmetric block cipher algorithm with a 128-bit key length. In the past, manufacturers could produce unsecured devices. Even today, Z-Wave networks can include secure and unsecured nodes for backward compatibility. However, security is now mandatory, so you can alleviate concerns by leveraging recent devices exclusively.

Z-Wave is an interesting choice for home automation or even smart building projects. Given its frequency band, it offers a greater range than Bluetooth and suffers from less interference than some alternatives. However, the proprietary control over some aspects of the technology is concerning.

Zigbee

Zigbee is a specification based on the IEEE 802.15.4 standard that defines a high-power communication protocol suite. Zigbee is under the stewardship of the Connectivity Standards Alliance (previously the Zigbee Alliance). The organization publishes application profiles that vendors can leverage to create interoperable products. The relationship between the IEEE 802.15.4 standard and Zigbee is comparable to the one between IEEE 802.11 and Wi-Fi.

Zigbee relies on the 2.4 GHz band in most jurisdictions worldwide. Specific devices also use 784 MHz in China, 868 MHz in Europe, and 915 MHz in the United States and Australia. The bulk of available devices sticks to the 2.4 GHz band, however. Data rates vary from 20 kbit/s (868 MHz) to 250 kbit/s (2.4 GHz).

Note   The name Zigbee refers to the dance that honeybees perform when they return to the beehive to inform their peers about the direction and distance of flower patches, water sources, or new nesting site locations.

Zigbee's focus on low power consumption limits the effective range between 10 and 100 meters in optimal conditions. Like Z-Wave, Zigbee devices can form mesh networks that will propagate data to their destination. The Zigbee standard defines three classes of devices:

- **Coordinator (ZC)**: The root node of the network. There is precisely one coordinator per Zigbee network. Coordinators can establish bridges to other networks.

- **Router (ZR)**: A node running application functions. Routers can propagate data to other devices.

- **End device (ZED)**: A leaf node that can only communicate with routers or the coordinator. Such devices typically spend extended periods in sleep mode, resulting in long battery life.

Only members of the Connectivity Standards Alliance can market products under the Zigbee name.

As with any wireless connectivity technology, there are security vulnerabilities in Zigbee. The protocol assumes that the keys used to operate and secure the network are stored securely and will not be transmitted unencrypted. Of course, devices with faulty security implementation or favoring usability and convenience over security could compromise their keys. Moreover, attackers could intercept packets and replay them to unsuspecting devices, tricking them into executing undesired operations. A possible mitigation measure would be distributing keys through out-of-band channels and rotating them regularly.

Like Z-Wave, Zigbee is a good choice for smart building projects and home automation. Given its range and flexibility, organizations also use it to track assets or monitor consumer energy consumption by integrating smart meters into the smart grid.

Network Design Considerations

Each networking technology I covered possesses characteristics that will influence how you design and deploy your network. However, a few relevant considerations apply independently of the technology. I will now review them.

Choice of Protocols

I covered several protocols in the first part of this book. Some require TCP/IP, some will be content with UDP, and others can leverage non-IP transports. Of course, the protocols you intend to use will influence the type of network you will deploy to support devices in the field. The cost of bandwidth and the expected reliability of the network selected will also influence the choice of protocols and vice versa.

Many field networks will not connect directly to the Internet, whether such a connection would be possible or not. An IoT gateway will bridge the local network with the rest of the infrastructure in such scenarios. A topology like this makes it possible to use, let's say, Bluetooth or Z-Wave locally to gather sensor data and send it to a Cloud MQTT broker over a cellular LTE or 5G link. The gateway, in this case, can pre-process the data to preserve the confidentiality or conserve bandwidth.

Confidentiality

Preserving the confidentiality of data, whether at rest (stored on the device) or in motion (during transmission), is a critical concern for any IoT solution. Naturally, several of the networking technologies I covered offer some level of encryption. But others, like Ethernet, do not. Therefore, planning for confidentiality is essential.

The most important advice I can give you is: Do not implement your own custom security layer! Even if you are an experienced developer, writing security-related code is a daunting task. The value of established security standards is that security researchers and large organizations pay attention to them, ensuring that new threats are discovered, discussed, and, eventually, addressed. If you build your own, the onus of watching and reacting to new threats will be on you.

If you follow my advice, what should you do then? First, if you use a protocol that relies on TCP/IP or UDP, then deploying TLS or DTLS is strongly recommended. You should stick to the most recent version of those standards supported by the protocol and operating system. In more concrete terms, this means to use TLS/DTLS 1.3 if possible and never use anything older than TLS/DTLS 1.2. An alternative strategy would be to perform payload encryption, although this would require you to determine if the specific encryption method used is still secure periodically.

Note   The version numbers for TLS and DTLS have been adjusted to be the same, starting with version 1.2.

Using a recent and supported version of TLS/DTLS is only the first step. You also need to configure an appropriate cipher suite on every device on the network and back-end resources in the Cloud or data center. The cipher suite you will select will impact battery life and bandwidth usage. In general, suites that offer a higher level of security have higher compute requirements due to more complex cryptographic calculations. Table 9-2 provides a view of the main characteristics of the three most appropriate authentication methods currently in use.

Table 9-2. *Common authentication methods (Credit: Kai Hudalla, Bosch.IO[5])*

	RSA	**ECDSA**	**PSK**
Minimal compute requirements	Cortex-M4	Cortex-M3	Cortex-M0
Certificate size	1–2 KB	<1 KB	0 B
Typical key length (bits)	2048	224–255	112
Full handshake size	4–7 KB	2–4 KB	<1 KB
Abbreviated handshake size	<500 B	<500 B	<500 B

Leveraging additional features in TLS/DTLS is also advisable. For example, leveraging the Ephemeral Diffie-Hellman (DHE) key exchange mode enables *forward secrecy* when supported by the cipher suite. A temporary key is generated for every connection when a key exchange uses the Ephemeral Diffie-Hellman algorithm. Those keys are not reused, which means that if the long-term private key of the back-end service is compromised, past exchanges are still confidential. Another useful feature

---

[5] This table has been adapted from a talk by Kai Hudalla at EclipseCon 2021 titled "Rock Solid Device Connectivity." You can watch it on YouTube: https://youtu.be/CfTIw1k2rGQ

is session resumption. TLS and DTLS handshakes take a lot of time and computing resources. Given this, session resumption provides a mechanism to reuse recently negotiated secrets across multiple connections. The abbreviated handshake that results is shorter and lighter on resources.

The keys and certificates that power most encryption solutions – not just TLS and DTLS – have a defined lifespan. You need to specify their expiration date at the time of creation. This means that any system relying on encryption will need to deal with the rotation (e.g., the periodic replacement) of credentials, such as keys and certificates. You could be tempted to avoid this issue by setting expiration dates very far in the future. Please do not do this, as it would create a massive security weakness in your infrastructure because keys and certificates are more likely to be compromised as time passes. Regularly replacing the credentials will reduce the length of the opportunity window for attackers. Of course, performing the rotation is unpleasant and potentially time-consuming, even with automation. In addition, the rotation can be challenging if you want to keep the keys private to the device itself since many constrained devices lack a good source of randomness, which is a requirement to create strong pairs of cryptographic keys. However, changing the keys and certificates used in your solution at a regular cadence will make it more secure.

Resiliency

IoT and edge computing solutions should assume the network will be unreliable and, at times, unavailable. This does not mean there is nothing you can do to make it more resilient. Let's have a look at a few best practices.

When devices are deployed in the real world, they are much more exposed to changes in network topology. Some of the connectivity options we explored are not based on TCP/IP and will route traffic through the neighboring nodes they can detect. Thus, they will automatically adapt to the addition or removal of nodes in their vicinity. In that case, building a resilient network means carefully positioning the nodes to ensure that the loss of a few will not compromise the connectivity of others. In the case of TCP/IP-based networks, the key thing is to rely on the Domain Name System (DNS) rather than IP addresses. This ensures the availability of services independently of the actual servers hosting them. Naturally, this strategy works better if you cache the results of DNS queries for a short time only. It will ensure that devices adapt to network topology or server availability changes more quickly.

Unless you work with constrained devices that need to maximize their battery life, leveraging the *keep alive* or *heartbeat* features in protocols will lead to more resilient network connections. For example, without them, a device could lose its connection to the carrier network without detecting it, making its TCP/IP session stale. Moreover, although session TLS/DTLS resumption leads to shorter handshakes, they can be avoided by keeping the connection alive. If you need to put devices to sleep to reduce power consumption, it could still be possible to reduce the burden on the devices by using protocol-specific features. For example, MQTT persistent sessions enable a client to reconnect to a broker without reestablishing its previous subscriptions; the session expiry mechanism in MQTT 5 ensures that the sessions of devices that do not reconnect in a timely fashion are purged. Such features do not alleviate the need to reestablish the underlying TCP/IP connection but streamline the process, nevertheless.

I recommend that you apply an exponential backoff strategy for reconnection attempts. Instead of retrying at a fixed interval, you will use an algorithm that incrementally increases the wait time between connection attempts. This technique will avoid overwhelming the back-end services with a flurry of requests, ensuring a smoother recovery once the services are back online.

Finally, another useful strategy is to use aggressive timeouts for connections and requests. Unless network latency is highly unpredictable, it does not make sense for devices to wait more than a few seconds for an attempt to get through. Such a "fail fast" strategy will also contribute to longer battery life, especially when used in conjunction with exponential backoff.

Redundant Connectivity

For mission-critical use cases, where network outages could provoke significant material damage or result in loss of human life, relying on a single network connection would be a single point of failure. Thus, many gateways and other IoT devices offer redundant network connections. For technologies relying on a cellular network such as LTE, 5G, LTE-M, or NB-IoT, this is typically achieved by adding a second standby SIM from a different network provider to the device. Since such providers typically operate on a subscription model where cost is determined by a plan, actual data consumption, or both, it is usually cost-effective to proceed that way. The redundant connection will likely see little traffic, hence controlling costs.

If your connectivity requirements are stringent, leveraging networks relying on completely different technologies and frequency bands is a sound, if more costly, strategy. Of course, you will then need to account for differences in effective range, data rates, latency, and software support.

Out-of-Band Management

There is a need to maximize uptime for devices deployed in remote or hard-to-reach locations in many enterprise scenarios. Since sending support personnel onsite is onerous and time-consuming, it is often cheaper to rely on a separate, out-of-band network connection to manage and troubleshoot devices. Such connections should use a different technology than the one handling the device's regular traffic. Of course, to keep costs under control, it can make sense to deploy out-of-band networks only at the level of IoT gateways or edge nodes.

In some cases, it makes sense to deploy specialized appliances that will connect to locally managed devices through their serial (UART) or Ethernet ports. Appliances like that enable connectivity when the network is down and when a device is turned off, in sleep mode, hibernating, or otherwise unavailable.

From a security perspective, segregating management traffic on a dedicated network can make the solution more resilient in case of attack. Unexpected or suspect activity on the out-of-band network is easier to spot, given its lower usage level than the primary connection.

Connectivity and Constrained Devices

Most of the time, support for network connectivity technologies is determined by the networking stack provided by the OS or RTOS you are using and the available drivers. Things are straightforward in the case of Linux; most radio modules currently available have direct support in the kernel tree or ship with the relevant drivers. Things are spottier in the case of RTOSes. Let's now look at using Wi-Fi and Bluetooth connections on the Zephyr RTOS.

Wi-Fi

There is only one generic sample application focused on Wi-Fi in the Zephyr code repository. It is described here: `https://docs.zephyrproject.org/latest/samples/net/wifi/README.html`.

The application will run on any board supporting Wi-Fi under Zephyr. It does not contain actual code; it rather shows how to configure an application for Wi-Fi and enable the `net_shell` and `net_wifi_shell` modules. Those modules provide a command-line interface you can use to run various network-related commands. For example, `net_shell` implements support for commands such as `ping` and `route`, while `net_wifi_shell` offers commands to scan for available Wi-Fi networks, connect to them, or disconnect from them.

I reproduced the output of the scan command from `net_wifi_shell`:

```
shell> wifi scan
Scan requested
shell>
Num  | SSID                  (len) | Chan | RSSI | Sec
1    | YoRHa no.2, Type B     8    | 1    | -93  | WPA/WPA2
2    | YoRHa no.9, Type S     8    | 6    | -89  | WPA/WPA2
3    | Nausicaä              13    | 11   | -73  | WPA/WPA2
4    | Forger Family          4    | 1    | -26  | WPA/WPA2
----------
Scan request done
```

Naturally, the `net_shell` and `net_wifi_shell` modules should not be used in production applications, given the resources they consume.

The Zephyr repository contains a good board-specific sample application focused on Wi-Fi named Espressif ESP32 Wi-Fi Station. As the name suggests, it supports the ESP32, ESP32-S2, and ESP32-C3 SOCs from Espressif. The code can be found at this location in the Zephyr GitHub code repository:

`https://github.com/zephyrproject-rtos/zephyr/blob/main/samples/boards/esp32/wifi_station`

The application is simple; it will connect to a network using an SSID and password specified statically in a configuration file. The following is a snippet from the main method showing how to connect:

```
wifi_config_t wifi_config = {
    .sta = {
        .ssid = CONFIG_ESP32_WIFI_SSID,
        .password = CONFIG_ESP32_WIFI_PASSWORD,
    },
};

esp_err_t ret = esp_wifi_set_mode(WIFI_MODE_STA);

ret |= esp_wifi_set_config(ESP_IF_WIFI_STA, &wifi_config);
ret |= esp_wifi_connect();
```

Since the constrained device plays the role of a Wi-Fi client, as opposed to an access point, the parameter passed to `esp_wifi_set_mode` is `WIFI_MODE_STA`.

The application relies on the ESP32 Wi-Fi driver, which is in a different location in the Zephyr repo:

`https://github.com/zephyrproject-rtos/zephyr/blob/main/drivers/wifi/esp32/src/esp_wifi_drv.c`

The driver, in turn, leverages the Espressif IoT Development Framework, which is hosted in a different repository in the Zephyr project's GitHub organization. The framework supports various features and peripherals found in the relevant SOCs. You can access the documentation for the stable version at this location: `https://docs.espressif.com/projects/esp-idf/en/stable/esp32/api-reference/index.html`.

Bluetooth

There is a great variety of Bluetooth samples in the Zephyr repository. The root location for them is

`https://github.com/zephyrproject-rtos/zephyr/blob/main/samples/bluetooth`

Mesh networking is an important capability of Bluetooth BLE technology. There are three samples focused on it. Here, I will focus on the one simply called mesh. Most of the code is contained in the main.c file, although there is a fun customization for the micro:bit that will use the board's LED display.

To add a Bluetooth device to a mesh, you need to provision it. Provisioning is a configuration and security process. It is managed by a *provisioner*, which can be another device or a software application. The provisioner discovers Bluetooth devices in the vicinity and invites them to join the network. If the device accepts, then the provisioner and itself exchange public keys. Each device then generates a session key that will secure the exchange of the provisioning data. This data includes the network key, a device key, a security parameter called the IV index shared among all the devices on the network, and the device's unicast address. Once the exchange concludes, the device is a node in the network.

The Bluetooth mesh specification integrates the notion of *models* to foster interoperability. The Bluetooth mesh glossary states that a model

> *[...] defines a set of States, State Transitions, State Bindings, Messages, and other associated behaviors. An Element within a Node must support one or more models, and it is the model or models that define the functionality that an Element has. There are a number of models that are defined by the Bluetooth SIG, and many of them are deliberately positioned as "generic" models, having potential utility within a wide range of device types.*

Note   For more information about Bluetooth mesh models, you should read this excellent white paper by Martin Woolley: Bluetooth Mesh Models – A Technical Overview.

The mesh Zephyr sample application defines several simple models. On boards featuring LEDs, a Generic OnOff Server model controls the state of the board's first LED over the mesh. On boards equipped with buttons, a Generic OnOff Client model sends OnOff messages to all nodes when a button is pressed. You can provision devices running the application the standard way or press a button to perform self-provisioning. A standard provisioner will need to supply an application key and bind it to both models. On the other hand, self-provisioning will bind a dummy application key to the models and use a random unicast address for the device.

Once you have provisioned the devices, pressing buttons will propagate OnOff messages in the mesh, turning the LED on or off upon reception.

Let's now dive into the code. The main function registers handlers for button presses and initializes the Bluetooth subsystem. Initialization is performed through a call to bt_enable. The only parameter is a reference to a callback function named bt_ready, as seen here:

```
err = bt_enable(bt_ready);
if (err) {
    printk("Bluetooth init failed (err %d)\n", err);
}
```

The critical line in the bt_ready function is this one:

```
err = bt_mesh_init(&prov, &comp);
```

The two parameters configure the provisioning process and the models, respectively. The provisioning configuration is contained in the struct reproduced here:

```
static const struct bt_mesh_prov prov = {
    .uuid = dev_uuid,
    .output_size = 4,
    .output_actions = BT_MESH_DISPLAY_NUMBER,
    .output_number = output_number,
    .complete = prov_complete,
    .reset = prov_reset,
};
```

Things are a bit less direct for the models. The comp struct, as shown here, refers to another struct called elements.

```
static const struct bt_mesh_comp comp = {
    .cid = BT_COMP_ID_LF,
    .elem = elements,
    .elem_count = ARRAY_SIZE(elements),
};
```

And elements, in turn, integrates another struct called models. This reflects that, in this case, the application declares a single element to hold all its models.

```
static struct bt_mesh_elem elements[] = {
    BT_MESH_ELEM(0, models, BT_MESH_MODEL_NONE),
};
```

In addition to the two custom models required by the application, the models struct uses standard models for configuration and health. The following is the declaration:

```
static struct bt_mesh_model models[] = {
    BT_MESH_MODEL_CFG_SRV,
    BT_MESH_MODEL_HEALTH_SRV(&health_srv, &health_pub),
    BT_MESH_MODEL(BT_MESH_MODEL_ID_GEN_ONOFF_SRV, gen_onoff_srv_op, NULL,
            NULL),
    BT_MESH_MODEL(BT_MESH_MODEL_ID_GEN_ONOFF_CLI, gen_onoff_cli_op, NULL,
            NULL),
};
```

As you can see, a list of possible operations is attached to the two custom models. Here is one of them:

```
static const struct bt_mesh_model_op gen_onoff_srv_op[] = {
    { OP_ONOFF_GET,       BT_MESH_LEN_EXACT(0), gen_onoff_get },
    { OP_ONOFF_SET,       BT_MESH_LEN_MIN(2),   gen_onoff_set },
    { OP_ONOFF_SET_UNACK, BT_MESH_LEN_MIN(2),   gen_onoff_set_unack },
    BT_MESH_MODEL_OP_END,
};
```

Each of those operations is tied to specific callback functions. For example, gen_onoff_get calls onoff_status_send.

```
static int onoff_status_send(struct bt_mesh_model *model,
                struct bt_mesh_msg_ctx *ctx)
{
    uint32_t remaining;

    BT_MESH_MODEL_BUF_DEFINE(buf, OP_ONOFF_STATUS, 3);
    bt_mesh_model_msg_init(&buf, OP_ONOFF_STATUS);

    remaining = k_ticks_to_ms_floor32(
                k_work_delayable_remaining_get(&onoff.work)) +
            onoff.transition_time;
```

```
...
    if (remaining) {
        net_buf_simple_add_u8(&buf, !onoff.val);
        net_buf_simple_add_u8(&buf, onoff.val);
        net_buf_simple_add_u8(&buf, model_time_encode(remaining));
    } else {
        net_buf_simple_add_u8(&buf, onoff.val);
    }

    return bt_mesh_model_send(model, ctx, &buf, NULL, NULL);
}
```

As you can see, most of the code in the function is used to build the payload, which is then transmitted by calling bt_mesh_model_send.

Working directly with Bluetooth networking involves writing a bit more code compared to protocols like LwM2M or MQTT. Using such protocols over a Bluetooth connection is possible by leveraging 6LoWPAN, allowing IPv6 packets to be transmitted over IEEE 802.15.4-based networks. However, this will probably consume more power than Bluetooth's native protocol.

CHAPTER 10

Operating Systems

Informatique : Alliance d'une science inexacte et d'une activité humaine faillible.

Computer science: Alliance of an inexact science and a fallible human activity.

—Luc Fayard

An operating system (OS) is software that manages a computer's hardware and software resources and provides shared services to other programs. In other words, it is an interface between hardware and software. Since the 1990s, most computers on the market run an operating system. However, this was not always the case. Early mainframes in the 1950s lacked one and could only run one program simultaneously. IBM was the first hardware maker to introduce the concept of a single operating system spanning an entire line of products with OS/360 in 1964 – although the company released three software variants due to storage limitations and the overall complexity of the project. In the 1970s and 1980s, many popular home microcomputers were used without an operating system. Nearly all of them shipped with a built-in BASIC interpreter that doubled as a command-line interface for the machine. Since floppy disk drives were expensive, cassette tape was the most common storage medium. Because those home computers had fixed configurations and could only run a single program at a time, an operating system was not strictly necessary.

This chapter will describe various operating systems and their feature sets. I will also take the time to introduce you to a few *bare metal* options, that is, ways to build applications that will run directly on the hardware without the support of an OS. Finally, I will show you how to get started as a developer with the Drogue IoT project and the Zephyr operating system.

© Frédéric Desbiens 2023
F. Desbiens, *Building Enterprise IoT Solutions with Eclipse IoT Technologies*,
https://doi.org/10.1007/978-1-4842-8882-5_10

Are Operating Systems Necessary?

Before going any further, let's ask ourselves if operating systems are necessary for enterprise IoT and edge computing deployments. The short version of the answer goes like this: for constrained devices performing simple tasks, maybe you can do without them. Alternatively, there are development frameworks called *unikernels* that ship the services usually provided by an operating system as libraries that are composed with the code of your application. If you find one fulfilling your requirements that can run on the hardware you plan to use, then going OS-less is possible.

From the previous paragraph and the very existence of this chapter, you probably deduced that there are plenty of enterprise IoT use cases where an OS or RTOS plays a useful role. Here is a list of features OSes offer that may make them relevant to your project:

- **APIs:** Application Programming Interfaces (APIs) hide the internal details of systems and often allow interoperability. When different operating systems implement the same APIs, applications or libraries are portable from one OS to another. A good example is the Portable Operating System Interface (POSIX), a family of standards maintained by the IEEE. The Zephyr RTOS implements a subset of the POSIX embedded profiles PSE51 and PSE52 and the BSD Sockets API.[1] RIOT OS also ships a POSIX wrapper providing implementations of semaphores, sockets, and threads.[2]

- **Hardware support:** OSes support many hardware components through device drivers. A driver is a program that controls a component connected to a computer through a bus or communication subsystem. They abstract the hardware details by implementing well-defined but operating system–specific interfaces exposed by the OS. Applications simply need to interact with these interfaces.

---

[1] See https://docs.zephyrproject.org/latest/services/portability/posix.html for details.
[2] https://doc.riot-os.org/group__posix.html

- **Memory management:** Desktop and server-class operating systems can provide a virtual address space to applications and handle page faults. On the other hand, RTOSes can allocate memory to applications and their own services statically or dynamically. The most advanced of them can create and manage memory domains and partitions. Domains ensure each user thread possesses its stack buffer, and partitions control access to specific memory zones. While most MCUs lack memory management units, a significant share of them sport memory protection hardware (MPU). MPUs monitor transactions such as instruction fetches and data accesses, which can trigger a fault exception when an access violation is detected.

- **Multitasking:** The capacity to execute concurrently multiple tasks (processes) is one of the big advantages of OSes. Delegating the control over concurrent processes and threads to the OS, as happens in preemptive multitasking, generally makes the solution more robust. This is especially true when features like memory protection are available. On the other hand, cooperative multitasking, where applications voluntarily cede computing time to others, is more fragile since a single badly written application can lead to system freezes.

- **Real-time requirements:** Real-time systems are designed to respond to events within defined, limited, and predictable timeframes. In other words, they feature predictable latency. Moreover, real-time systems explicitly manage resources in real time. Getting real-time latency and resource management right is complicated. Therefore, it makes sense to delegate those responsibilities to the OS.

- **Shared services:** Networking, security, storage access, filesystem management, and user interfaces are just some of the services operating systems offer. Leveraging those services makes applications smaller and simpler, helping you focus on the business problems you are trying to solve.

Overall, operating systems have a lot to offer. Nevertheless, there are viable alternatives that we will explore together later in this chapter.

Real-Time Operating Systems

There is a bewildering diversity of RTOSes in the market right now. As of writing, the list compiled by Wikipedia listed over 150 active RTOSes, both open source and proprietary. I decided to cover in detail three open source options that are actively maintained, have a strong ecosystem, support a wide variety of hardware, and offer robust toolkits to developers. They are presented in alphabetical order. Naturally, there are alternatives that I do not cover that could be a great fit for your project. Please take the time to explore possibilities before making your choice.

Arm Mbed OS

Given how popular microcontrollers based on the Cortex-M architecture are, you are probably not surprised to see I cover Arm's RTOS here. Released for the first time in 2009, Arm Mbed OS is an open source operating system for IoT devices based on Cortex-M MCUs. The project is developed in the open on GitHub[3] and managed by Arm and its hardware partners. The code is licensed under the Apache License v2.0. This makes it a good fit for open and commercial projects alike. At the time of writing, the current version of Mbed OS was v6.15.1, released in November 2021.

Mbed OS' architecture is highly modular, as shown in Figure 10-1.

---

[3] https://github.com/ARMmbed/mbed-os

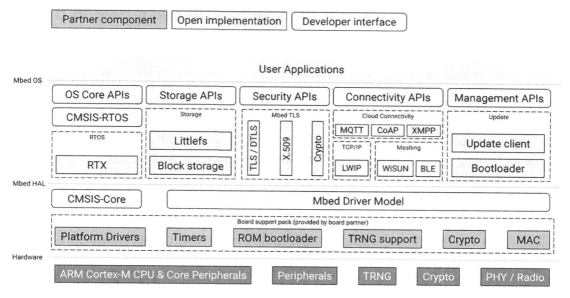

Figure 10-1. *Arm Mbed OS 6 architecture (Credit: Arm[4])*

Please note that this diagram is a simplification of the one provided by Arm. Specifically, I removed the details pertaining to Trusted-Firmware Cortex-M technologies. Those technologies, along with the TrustZone instructions found in some Cortex-M cores, are commonly used to create Trusted Execution Environment (TEE) instances that protect the confidentiality and integrity of the code and data loaded inside.

Naturally, given Arm's focus on hardware intellectual property licensing, the features of Mbed are highly dependent on what Cortex-M cores offer. However, other components, such as the wireless radios and the True Random Number Generator (TRNG), also play a crucial part. Since MCUs based on Cortex-M can greatly vary in layout and capabilities, support for a particular board is usually provided through a *board support pack* maintained by the hardware vendor.

Note   The TRNG comprises several sub-blocks in the integrated circuit making up the MCU that collaborate to generate cryptographically secure random numbers.

---

[4] The original figure can be found on this page: https://os.mbed.com/docs/mbed-os/v6.15/introduction/architecture.html

Mbed OS exposes two sets of APIs (named developer interfaces in Figure 10-1) to abstract the hardware details from the OS's upper layers. The first one is the Common Microcontroller Software Interface Standard (CMSIS): a vendor-independent abstraction layer for microcontrollers based on Arm Cortex cores. The second is the Mbed driver model that you can leverage when writing new drivers. Together, those two APIs make the Mbed OS hardware abstraction layer (HAL). This foundation facilitates writing applications since it provides libraries and drivers for standard MCU peripherals, such as I²C, Bus, and SPI. Because of the HAL, you can write applications in C and C++ that will run on any Mbed OS–enabled board.

Mbed OS can be configured to conform to two distinct profiles. The *full* profile is an RTOS that includes Keil RTX and all the relevant RTOS APIs. This profile supports deterministic, multithreaded, real-time applications. Drivers and applications can leverage threads, semaphores, mutexes, and other RTOS features. The other profile is called *bare metal* and focuses on minimizing code size. It only implements a subset of the Mbed OS RTOS APIs useful in nonthreaded applications, such as semaphores and tickers. Consequently, all activities are polled or driven by interrupts under the bare metal profile.

Note   Kiel is a subsidiary of Arm based in Germany. It provides a broad range of development tools (compilers, assemblers, debuggers, IDEs) supporting Arm hardware and evaluation boards.

The remainder of the high-level Mbed OS APIs belong to the four following categories:

- **Storage**: Mbed OS supports a wide array of filesystems. One is LittleFS, a high-integrity embedded filesystem optimized to run on limited amounts of RAM. It is resilient to power losses and ensures data integrity. Support for the popular FAT filesystem ensures interoperability with most other operating systems. Mbed can also manage block storage, enabling the creation of raw storage volumes.

- **Security**: Out of the box, Mbed OS supports cryptography and can manage X.509 certificates. It also provides TLS and DTLS implementations that enable encrypted communications for applications.

- **Connectivity**: Mbed OS supports TCP/IP through a port of the lwIP project, a small independent implementation of the TCP/IP protocol suite. The OS can work with two mesh networking technologies: Bluetooth BLE and Wi-SUN. The Wi-SUN Field Area Network (FAN) is based on open standards from IEEE 802, IETF, ANSI/TIA, and ETSI. The Mbed OS Wi-SUN stack is built on 6LoWPAN, which relies on IEEE 802.15.4. Mbed OS also supports NFC (for contactless payments and access control) and LoRaWAN.

- **Management**: Mbed OS ships with management APIs that enable over-the-air firmware updates. To leverage this feature, you need to use the Mbed OS–provided bootloader on your device.

The tooling ecosystem for Mbed OS is extensive. You can build applications using Arm's commercial C/C++ compiler (Arm Compiler) or the GNU Arm Embedded toolchain. In terms of development tools, you have three main options at your disposal:

- **Arm Keil Studio:**[5] A free-to-use, browser-based IDE. It supports Mbed OS projects and other CMSIS-compliant targets such as FreeRTOS. Keil Studio is the successor to the Arm Mbed Online Compiler, which was deprecated at the end of 2021.

- **Mbed CLI 2:**[6] A command-line environment using the Ninja build system and CMake to create the build environment and manage the build process independently of the compiler.

- **Mbed Studio:**[7] A desktop IDE for Windows, Linux, and macOS.

It is also possible to work on Mbed OS projects using several third-party development tools.[8]

---

[5] https://developer.arm.com/documentation/102497/1-5/Arm-Keil-Studio

[6] https://os.mbed.com/docs/mbed-os/v6.15/build-tools/mbed-cli-2.html

[7] https://os.mbed.com/docs/mbed-studio/current/introduction/index.html

[8] https://os.mbed.com/docs/mbed-os/v6.15/build-tools/third-party-build-tools.html

---

Note   Both Keil Studio and Mbed Studio integrate technology from the
Eclipse Theia project. Theia provides an extensible platform to build Cloud and
desktop IDEs.

Mbed OS is a strong choice if you want to leverage MCUs based on the Cortex-M
line. However, Arm remains firmly in control of the project. The lack of vendor-neutral
governance around the OS is certainly a concern from a community perspective.

FreeRTOS

FreeRTOS has been for a long time the dominant RTOS in the market for constrained
devices, open source or not. In the 2021 edition of the Eclipse IoT and Edge Developer
Survey, close to 30% of respondents stated they are using it. Part of the reason for its
popularity is its longevity. The FreeRTOS kernel was created by Richard Barry around
2003 and was later maintained by Barry's company, Real Time Engineers. In 2017, the
stewardship of the project was passed to Amazon Web Services (AWS). Barry is still
involved as an AWS employee. Most of FreeRTOS is written in C, although assembly
language is used in architecture-specific scheduler routines. At the time of writing, the
latest version of the kernel was v10.4.6, released in November 2021. The latest version of
the full OS package, including sample projects and additional libraries, was v202112.00.
The kernel and all other components maintained by the FreeRTOS team are available
under the MIT license.

FreeRTOS enjoys broad hardware support. At the time of writing, there were 24
hardware partners listed on the FreeRTOS site, and the open source community
supports several other platforms.[9] The team considers each architecture and compiler
combination to be a separate port. The ports belong to one of the four following
categories:

- Ports created and maintained by the FreeRTOS team (officially
  supported ports)

- Ports contributed by a partner and maintained by the FreeRTOS team

---

[9] See `www.freertos.org/RTOS_ports.html` for up-to-date details.

- Ports contributed by a partner and maintained by the partner

- Community-supported ports

Those ports are maintained in separate Git repositories but are exposed as submodules in the main FreeRTOS repository. Long-Term Support (LTS) releases only contain ports the FreeRTOS team maintains. Moreover, commercial support is available, but only for officially supported ports.

The FreeRTOS team does not provide ship development tools directly. Most of the demos in the project's main Git repository are configured to use the Eclipse IDE and specifically the Eclipse CDT (C/C++ Development Tooling) edition. Using the specialized Eclipse Embedded CDT edition, which comes with plugins to create, build, and manage applications targeting Arm and RISC-V platforms, is also possible. Many hardware providers also ship their own IDEs based on Eclipse IDE or Eclipse Theia. Naturally, given the simplicity of the code base, most IDEs supporting the C language can be used.

Compared to the other OSes I cover in this section, the structure of FreeRTOS is very simple. The team maintains three specific sets of libraries besides the kernel itself. Let's have a closer look at those four components.

FreeRTOS Kernel

The FreeRTOS kernel requires minimal resources. A typical binary image will typically weigh between 6 and 12 KiB. The core implementation is simple since it is contained in only three C files.

FreeRTOS provides primitives to manipulate tasks, mutexes, semaphores, and software timers. Developers can specify thread priorities. The kernel supports static memory allocations. Dynamic allocation is also possible, and the kernel ships with five sample memory management schemes. Those five schemes are as follows:

- **heap_1**: The kernel will allocate memory, but it cannot be freed.

- **heap_2**: The kernel will allocate and free memory but will not merge freed memory blocks with adjacent free blocks. This scheme is a legacy one, and heap_4 is preferred.

- **heap_3**: The kernel will rely on the malloc() and free() implementations exposed by the standard C library provided with your compiler. Those calls are wrapped for thread safety.

- **heap_4**: The kernel will allocate and free memory. It will also merge freed memory blocks with adjacent free blocks to avoid fragmentation.

- **heap_5**: The kernel will manage memory like in heap_4 and can spread the heap across noncontiguous areas.

Of course, you could also provide your own memory management implementation. The choice between static and dynamic memory allocation will come down to whether you favor simplicity and, potentially, lower memory usage over control of memory locations.

As you can imagine, the functional scope of the FreeRTOS kernel is very small, given its size. Fortunately, the team also provides several libraries implementing useful connectivity and security features.

FreeRTOS+ Libraries

The libraries in this set have a dependency on the FreeRTOS kernel. At the time of writing, there were two libraries in the FreeRTOS+ category:

- **FreeRTOS+TCP:** A complete TCP/IP stack exposing a re-entrant and thread-safe API based on the widespread Berkeley sockets interface. It includes DHCP and DNS, among other features. An optional callback interface is also available. The code size with all features enabled varies between 20.1 and 34.9 KiB, depending on the compiler optimizations selected.

- **FreeRTOS+CLI:** A small and extensible framework to add a command-line interface (CLI) to your FreeRTOS applications.

A third library, FreeRTOS+IO, provided a POSIX-like API to peripheral driver libraries but has been deprecated.

FreeRTOS Core Libraries

The libraries in this category implement connectivity and security features based on open standards. They depend only on the standard C libraries and, as such, can be used with other kernels than the FreeRTOS one. Here is the list of libraries in this category:

- **coreMQTT:** A lightweight MQTT client. The library implements MQTT 3.1, and all QoS levels are available. It also supports TLS for encrypted communications. The code size varies between 5.7 and 6.9 KiB, depending on the optimizations.

- **coreMQTT Agent:** A library that expands coreMQTT by adding a thread-safe API. You can use it to create a dedicated task to handle MQTT traffic in your application; other tasks will not be allowed to use the coreMQTT API. CoreMQTT Agent includes the code of coreMQTT. Code size is between 7.4 and 8.9 KiB.

- **coreHTTP:** A partial client implementation of the HTTP protocol. The library exposes a synchronous API that you can use to serialize headers, send a request to a remote server, and process the response. The library works with TLS and can be used for calling REST web services. Code size varies between 15.6 and 18.9 KiB.

- **coreSNTP:** A client implementation of the Simple Network Time Protocol (SNTP). The library implements v4 of SNTP as defined in RFC 4330. Synchronizing the clock of your constrained devices ensures that the times reported by your applications are accurate and can alleviate the lack of a real-time clock in some devices. SNTP is a subset of the Network Time Protocol (NTP) found on personal computers and servers; it requires fewer memory and compute resources than NTP. Code size falls between 2.0 and 2.5 KiB.

- **Transport Interface:** A way to decouple protocol implementations from the underlying network drivers. Implementations of the interface contain function pointers and context data for sending and receiving over a network connection. FreeRTOS provides sample implementations, and you can provide your own. Since they rely on the Transport Interface, coreMQTT and coreHTTP do not have a dependency on any specific TCP/IP stack, including FreeRTOS+TCP.

- **coreJSON:** A JSON parser that complies with the JSON data interchange syntax standard (ECMA-404). The library supports key lookups and relies on an internal stack to track nested structures. Code size is between 2.4 and 2.9 KiB.

- **corePKCS #11:** A hardware-independent implementation of the PKCS#11 API that specifies an API for devices storing cryptographic information and performing cryptographic functions. Usually, the makers of cryptographic hardware such as Trusted Platform Modules (TPM) and Hardware Security Modules (HSM) provide a PKCS#11 implementation with their products. The corePKCS11 library lets you quickly prototype and build applications before switching to a vendor-specific PKCS#11 library in the production code.

- **FreeRTOS Cellular Interface Library:** A library that encapsulates the TCP/IP stack for three popular cellular modems and exposes a uniform API. The three supported modems are the Quectel BG96, the Sierra Wireless HL7802, and the U-Blox Sara-R4.

FreeRTOS for AWS IoT

The libraries in this category provide clients for various AWS-specific IoT Cloud services. The features cover over-the-air updates (AWS IoT OTA), digital twins (AWS IoT Device Shadow), task notifications (AWS IoT Jobs), security metrics reporting (AWS IoT Device Defender), device provisioning (AWS IoT Fleet Provisioning), and digital signatures (AWS Signature Version 4). The libraries depend only on the standard C libraries and can be used with other kernels than the FreeRTOS one.

The Lowdown

Overall, FreeRTOS is a flexible and very small footprint operating system. Its broad support in the industry means a wealth of compatible libraries and several hardware suppliers to choose from. The lack of vendor-neutral governance around the project is a concern, however. Additionally, the tight integration with the AWS ecosystem is both a productivity boost and a potential long-term liability. Although the officially supported ports and libraries are of known origin and are not burdened by intellectual property issues, the same cannot necessarily be said for partner or community-maintained ports. The small functional footprint also means you will need to source third-party libraries to support specific hardware or features, which could lead to maintenance issues.

Zephyr

The last RTOS I will cover in detail is Zephyr. The roots of Zephyr go back to Virtuoso RTOS from Eonic Systems, a Belgian software company that Wind River Systems bought in 2001. Wind River renamed Virtuoso as Rocket and made it available as open source in 2015. Wind River contributed the project to the Linux Foundation under the name Zephyr in 2016. Zephyr is available under the Apache v2.0 license and is developed on GitHub.[10]

Zephyr's feature set is extensive. It includes a small monolithic kernel, a set of connectivity protocol implementations (including MQTT, CoAP, and LwM2M), a virtual filesystem interface supporting several types of flash storage, and built-in mechanisms for device management and software updates. Bluetooth BLE is also fully supported, including mesh networking. Zephyr is highly configurable and modular; you can incorporate only the features you need. Consequently, the runtime memory usage can be as low as 8 KiB. Figure 10-2 summarizes Zephyr's architecture.

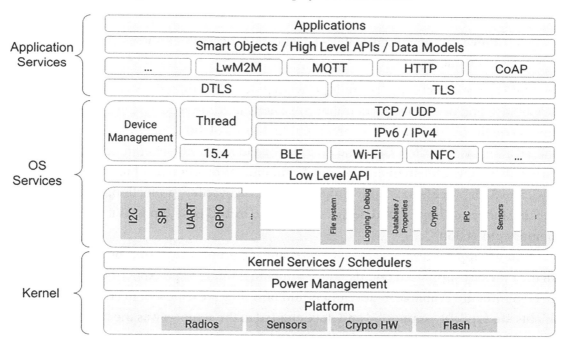

Figure 10-2. *Zephyr RTOS architecture (Credit: Linux Foundation[11])*

---

[10] https://github.com/zephyrproject-rtos/zephyr

[11] This is a slightly altered version of the diagram found in the official Zephyr documentation. It can be found here: https://docs.zephyrproject.org/latest/security/security-overview.html

One element not readily apparent in Figure 10-2 is the suite of kernel services. Here is the complete list of services:

- **Multithreading:** Provides you with cooperative, priority-based, nonpreemptive, and preemptive threads. Round-robin time slicing is optional. The service also provides a POSIX-compatible API.

- **Interrupts:** Implements registration of interrupt handlers at compile time.

- **Memory allocation:** Offers dynamic memory allocation. Memory blocks can be freed whether their size is fixed or variable.

- **Interthread synchronization:** Provides binary semaphores, counting semaphores, and mutex semaphores.

- **Interthread data passing:** Implements message queue and byte streams.

- **Power management:** Offers advanced idling features and a tickless idling mode.

The Zephyr kernel supports multiple ways to schedule threads. At a high level, both cooperative and preemptive scheduling are available. Time slicing, where processing time is equally divided among all preemptible threads of the same priority, is also offered along with Earliest Deadline First (EDF) and Meta IRQ scheduling. Moreover, the kernel offers powerful architecture-specific memory protection mechanisms. On x86, ARC, and Arm processors, Zephyr implements stack overflow protection, permission tracking for kernel objects and device drivers, and thread isolation with thread-level memory protection.

Zephyr supports a wide variety of hardware, including most Arm Cortex-M cores. All the officially supported boards[12] have a documentation page explaining which of their features are available to developers and how they are exposed. In addition, several shields are explicitly supported and documented in the same way as the boards.

---

[12] See https://docs.zephyrproject.org/latest/boards/index.html for the list.

Note   A shield is an expansion board fitted on the top of a microcontroller board. They are equivalent to HATs (Hardware Attached on Top) in the Raspberry Pi ecosystem.

From a development tools perspective, the Zephyr team maintains a software development kit providing complete toolchains for each of the supported architectures. The main dependencies for the SDK are CMake, Python, and the Devicetree compiler. All three support multiple platforms; the SDK can be installed on Linux, macOS, and Windows. In the case of Windows, using the Chocolatey package manager is recommended; you could also leverage the Windows Subsystem for Linux (WSL) feature available in Windows 10 and higher, but at the time of writing, application flashing was not possible in that environment. The Zephyr team does not ship an IDE, but several board makers support Zephyr in their IDEs derived from Eclipse CDT. Using PlatformIO and its namesake IDE is also a possibility.

An interesting feature of Zephyr is West, its companion meta-tool. West provides developers the ability to work with multiple Git repositories. It also supports developer workflows through a user-friendly command-line interface. West is compatible with CMake, which means you can perform operations such as building, flashing, and debugging with either tool. The use of West is optional, although bypassing it will probably not be convenient for you.

Zephyr is a well-documented RTOS with a comprehensive feature set and a mature code base. The vendor-neutral stewardship of the project by the Linux Foundation ensures that no one organization can wrestle away the project from the open source community. Zephyr offers a stable target to application developers since it publishes Long-Term Support (LTS) releases on a defined schedule. The project's governance is transparent, and the team takes security very seriously. A long-term goal of the project is to maintain a secure branch to create a certifiable system or at least a subset of certifiable submodules.

For all these reasons and given that the Eclipse Foundation is a member of the Zephyr project, I based most of the constrained device code samples in this book on Zephyr. However, Arm Mbed and FreeRTOS are strong alternatives. There are also other open source RTOSes worthy of your attention that I did not have the time to cover, like Contiki and RIOT OS. In the end, you should pick the option that delivers the best balance of features and hardware support for your project.

Bare Metal Options

While RTOSes have a lot to offer, they are probably overkill for simple use cases where the constrained device performs simple tasks one at a time. In that context, the overhead of the RTOS may outweigh the benefits. Moreover, there is a growing number of frameworks that deliver most of the features offered by RTOSes and produce programs that are run directly on the hardware without the involvement of an operating system. In this section, I will review a few options.

Arduino

Out of the box, Arduino boards execute programs directly, with or without the assistance of a bootloader.[13] Such programs are called *sketches* in Arduino parlance. The bootloader's role is to allow you to upload sketches to the board without needing a hardware programmer, a specialized device that can burn the software on a board. You can operate the board without a bootloader if you have a programmer and need all the Arduino's resources for your program.

Arduino sketches are written in a simplified version of C++. At compile time, a pre-processor converts the sketch into standard C++ code. The object file is linked against the standard Arduino libraries. The result is a single hex file that will be executed directly by the board. This approach is powerful and modular due to the extensive collection of optional libraries made available by Arduino and the wider community. The libraries cover various domains such as communications, connectivity, data processing, storage, and device control. The Arduino CLI command-line tool and Arduino IDE make it easy to install the core support files for the board of your choice and the libraries you need.

The Arduino is a fantastic example of a streamlined open source environment targeted at simple use cases. That said, several RTOSes can run on Arduino boards, including FreeRTOS.

---

[13] For more information about the Arduino bootloader, see `https://docs.arduino.cc/hacking/software/Bootloader`

Drogue IoT

If you are a fan of the Rust programming language (and if you are not one, you should at least entertain the thought), Drogue IoT provides everything you need to build secure and efficient firmware for constrained devices. Red Hat has created the project; it is not an Eclipse Foundation project, but several of its contributors play a role in the Eclipse IoT and Edge ecosystem. You can find the Drogue IoT website at www.drogue.io/

The usage of Rust has been growing in the IoT and embedded space in the last few years. This is partly due to the many safety features in the language. The Rust type system prevents data races at compile time and can perform static checks. These static checks can be leveraged, for example, to enforce the proper configuration of I/O interfaces and ensure that operations are performed only on properly configured peripherals. The checks can also implement access control to ensure that only certain parts of a program can modify the peripheral state. In addition, Rust development tools are available on all major operating systems, can target multiple hardware platforms, and are not tied to any specific project or framework. Developers can pick from an extensive collection of open source libraries published on crates.io, the Rust equivalent to Maven Central. Those libraries are called *crates*, unsurprisingly.

Building device firmware in Rust does not require vendor-specific SDKs or RTOS-specific tooling. However, the community of Rust embedded developers standardized a few concepts. Those concepts are as follows:

- **Peripheral Access Crate (PAC):** Libraries that allow access to chip-specific registers. Static checks ensure developers are using the peripherals appropriately.

- **Hardware Abstraction Layer (HAL):** Libraries that build upon the PAC of certain chips and provide abstractions of common features and buses such as I²C, SPI, and UART. HALs often work across members of a chip family.

- **Embedded HAL:** Libraries that are a foundation for building platform-agnostic drivers. They define interfaces that are then implemented in HALs.

There are two major components in Drogue IoT: Drogue Device and Drogue Cloud. Drogue Device is a distribution of libraries and drivers for building IoT applications. Behind the covers, it relies on the Embassy project. Embassy is an asynchronous programming framework comprising an executor and a Hardware Access Layer (HAL). The executor is a scheduler. It executes a set of tasks defined at startup time, although you can add more later. The HAL provides an API encapsulating access to peripherals such as USART, UART, I²C, SPI, CAN, and USB. Where it makes sense, Embassy APIs are offered in both synchronous and asynchronous flavors. Embassy represents an alternative to the Real-Time Interrupt-driven Concurrency (RTIC) library, a runtime providing a task scheduler and many of the core features typically found in an RTOS.

There are several advantages to asynchronous Rust over the traditional synchronous approach. First, it produces power and resource-efficient applications since async tasks can be completed by raising interrupts. Second, it simplifies developers' lives by removing the need to implement state machines in complex scenarios such as using multiple DMA channels at once. The Rust compiler will generate the state machine from the "linear" Rust async code implementing the logic. Finally, asynchronous Rust tasks can benefit from a perfectly sized stack through a construct called *futures*, representing "[…] single eventual values produced by asynchronous computations."[14] Segmenting the future state and allocating memory only when needed is possible. Once again, this results in more efficient code.

Embassy is a task-based framework. The Drogue Device programming model supports that and also provides an actor framework. An actor system isolates states in narrow scopes; actors represent the boundary of state usage. In Drogue Device, actors make it easier to build concurrent systems using message passing. Each actor possesses a unique address. Actors are decoupled from one another and process only one message at once. Specifically, they process messages by leveraging Rust's async/await primitives. Consequently, events are always processed in order.

Drogue Device ships with optional board support packages that provide boilerplate for the most used peripherals. The project also includes drivers, actors, and examples for Wi-Fi, LoRaWAN, and Bluetooth BLE combined with various sensors. You can also leverage third-party drivers or write your own. At the time of writing, Drogue Device could run on any hardware supported by Embassy. The list includes the STM32 series

---

[14] https://docs.rs/futures/latest/futures/

from STMicroelectronics and the Nordic Semiconductor nRF52, nRF53, and nRF91 series. An example of a well-supported and fully featured board is the micro:bit version 2.

Drogue Device is a powerful framework for building Rust applications targeting embedded and IoT constrained devices. The language provides a low-footprint runtime, memory safety, and thread safety, a winning combination given the target use cases. As for the companion Drogue Cloud platform, it exposes HTTP, MQTT, and CoAP endpoint to devices and performs protocol normalization through Cloud Events and Knative eventing. The integration between Drogue Device and Drogue Cloud means that Drogue Cloud can deliver effective device management, including credentials and configuration properties.

Espressif IDF (ESP-IDF)

The Espressif IoT Development Framework (ESP-IDF) is another great example of a bare metal framework for constrained devices. Out of the box, it supports several boards equipped with Espressif's ESP32 MCU, which features Wi-Fi, Bluetooth, and two Xtensa LX6 32-bit processor cores. The framework is published under the Apache v2.0 license, and you can find the source code in this GitHub repository: `https://github.com/espressif/esp-idf`. The documentation is available at `https://docs.espressif.com/projects/esp-idf/en/latest/esp32/index.html`.

> **Note**   Xtensa processors are based on a 32-bit RISC architecture distinct from Arm and RISC-V. Given the ESP32's popularity, one could surmise that Xtensa cores are among the most widely adopted alternatives to those two.

ESP-IDF possesses a comprehensive feature set. There are built-in clients for HTTP and MQTT, and it is possible to instantiate HTTP and HTTPS servers. Legacy protocols such as Modbus are also supported. Since the ESP32 is an MCU, you can leverage several peripheral interfaces such as GPIO, I²C, SPI, and UART. The Storage API supports many device types, including SD cards and flash memory.

One particularity of ESP-IDF is that it relies on a modified version of FreeRTOS as its scheduler. Espressif added support for symmetrical multiprocessing (SMP) to it since the ESP32 is a dual-core chip. Consequently, there are a few differences between plain

FreeRTOS code and ESP-IDF code. Given all this, you could wonder if ESP-IDF is really a bare metal framework. I think it qualifies due to the tight integration between the components and the rich feature set.

There are two ways to install the tooling provided by Espressif: through an IDE or manually. The team ships a plugin for the desktop Eclipse IDE and a Microsoft Visual Studio Code extension. You can use Linux, macOS, or Microsoft Windows as your operating system.

ESP-IDF is, without a doubt, a powerful and flexible framework. Naturally, the fact it is tied to a specific CPU architecture limits its appeal, although the ESP32 is worth your consideration if you are in the market for a 32-bit MCU.

Edge Nodes: The Realm of Linux

Up to now, I have exclusively discussed operating systems in constrained devices. Time to have a look at the landscape for edge nodes, then. The 2021 edition of the Eclipse IoT and Edge Developer Survey found that Linux is the top choice for both gateways and edge servers, with Microsoft Windows in second place. This is no surprise. Linux's modularity and extensive hardware support make it a strong choice for various workloads. Even offerings such as Microsoft's Azure Sphere, a tightly integrated platform combining a secure MCU with a custom software stack, rely on the Linux kernel.

In the same survey, we also asked respondents what workloads they are running at the edge and how they are packaging the edge services they deploy. The workloads are quite diversified; we will revisit them in the chapter focused on edge computing. As for service packaging, containers were the clear first choice. Such workloads are strongly correlated with the use of Linux.

One point I did not cover yet is the use of Linux in constrained devices. After all, Linux is widespread in embedded systems, and it is possible to boot the kernel on any device sporting a suitable 32-bit processor and at least 4 megabytes of memory. Specifically, the processor core must be equipped with a hardware memory management unit (MMU). Plenty of systems on a chip (SOCs) fulfill those requirements and are suitable for embedded use. However, such SOCs consume much more power than most MCUs. Ultimately, whether Linux is suitable for use in constrained devices is highly dependent on how you define such devices in the first place.

Of course, the standard Linux kernel is optimized for throughput; it cannot provide the kind of predictable latencies developers expect from an RTOS out of the box. However, activating specific pre-emption models in the mainline kernel tree allows high-priority tasks to pre-empt currently running processes. Moreover, the integration of the popular PREEMPT_RT patchset into the mainline is in progress. PREEMP_RT makes the Linux kernel fully preemptible, and the runtime behavior is very close to a pure RTOS. That said, not all Linux device drivers are compatible, which can reduce its attractiveness to some.

Getting Started with Zephyr

The Zephyr team built its SDK to support the three main operating systems used by developers: Linux, macOS, and Windows. All the dependencies are cross-platform. However, you must pay attention to the minimum version numbers supported. Table 10-1 lists the minimum version numbers for the three main dependencies of the SDK.

Table 10-1. *Minimal supported versions for the main Zephyr SDK dependencies*

Dependency	Minimal version
CMake	3.20.0
Python	3.6
Devicetree compiler	1.4.6

Getting a recent version of CMake is critical. On Ubuntu, you can rely on the APT repository made available by Kitware, the main contributor to the project. You can also download binary distributions from their website for other distributions or operating systems.

Linux

To install the SDK dependencies on Ubuntu, execute this command:

```
sudo apt-get install --no-install-recommends git cmake \
```

```
ninja-build gperf ccache dfu-util device-tree-compiler wget \
python3-dev python3-pip python3-setuptools python3-tk \
python3-wheel xz-utils file libpython3.8-dev \
make gcc gcc-multilib g++-multilib libsdl2-dev
```

For Fedora, the command looks like this:

```
sudo dnf group install "Development Tools" \
  "C Development Tools and Libraries"
sudo dnf install git cmake ninja-build gperf ccache dfu-util \
  dtc wget python3-pip python3-tkinter xz file glibc-devel.i686 \
  libstdc++-devel.i686 python38 SDL2-devel
```

Once the package manager is done, you must prepare Python since West relies on it. It is a best practice to leverage Python virtual environments when working on Zephyr projects. You can create and activate one like this:

```
python3 -m venv ~/zephyrproject/.venv
source ~/zephyrproject/.venv/bin/activate
```

Substitute the name of your choice for `zephyrproject`. Remember to activate the virtual environment every time you start a new shell instance to work on the code.

Once this is done, you must add the West tool to your Python environment. This is the relevant command:

```
pip install west
```

The next step is to get the Zephyr source code. You could have cloned the repository, but West can also do this.

```
west init ~/zephyrproject
cd ~/zephyrproject
west update
```

West can configure the environment to allow CMake to load the Zephyr boilerplate code required by the build process. Run this command:

```
west zephyr-export
```

Before continuing, you need to install a few additional Python dependencies. They are found in `zephyr/scripts/requirements.txt` in your project's folder.

```
pip install -r ~/zephyrproject/zephyr/scripts/requirements.txt
```

After that, time to download and install the SDK itself.

```
wget https://github.com/zephyrproject-rtos/sdk-ng/releases/download/
v0.14.2/zephyr-sdk-0.14.2_linux-x86_64.tar.gz
wget -O - https://github.com/zephyrproject-rtos/sdk-ng/releases/download/
v0.14.2/sha256.sum | shasum --check --ignore-missing
```

You can download a specific SDK version by replacing 0.14.2 in the preceding commands with the correct version number. You can access the list of releases here: https://github.com/zephyrproject-rtos/sdk-ng/releases.

From the folder where you downloaded the SDK, you can unpack and run the installer by executing these commands:

```
tar xvf zephyr-sdk-0.14.2_linux-x86_64.tar.gz
cd zephyr-sdk-0.14.2
./setup.sh
```

To avoid problems, please unpack the SDK in one of the following locations:

- $HOME

- $HOME/.local

- $HOME/.local/opt

- $HOME/bin

- /opt

- /usr/local

By the way, you need to download the SDK for the architecture of your development machine. Get the x86_64 version for a workstation with an Intel or AMD processor and aarch64 for one with a processor using the Arm ISA.

macOS

The Zephyr team assumes you are using the Brew package manager on macOS. To install the dependencies, execute this command:

```
brew install cmake ninja gperf python3 ccache qemu dtc wget
```

The rest of the process uses the same commands as on Linux; please refer to the previous section. For the SDK, select the x86_64 version for an Intel Mac and aarch64 for an Apple Silicon Mac (M1 and successors).

Windows

The process on Windows follows a similar flow, but the commands are, of course, slightly different. The instructions provided by the Zephyr team use the Chocolatey package manager. I assume here you already set it up.

Open an elevated command prompt (prompt with Administrator rights) and execute the following commands:

```
choco install cmake --installargs 'ADD_CMAKE_TO_PATH=System'
choco install ninja gperf python git dtc-msys2 wget unzip
```

You can close the elevated shell once you are done; you can perform the rest of the steps under a standard shell. To create and activate a virtual Python environment, execute the following commands:

```
cd %HOMEPATH%
python3 -m venv zephyrproject\.venv
zephyrproject\.venv\Scripts\activate.bat
```

Replace zephyrproject with a name of your choosing. If you use PowerShell rather than the standard command-line interpreter (cmd.exe), then the last command should rather be

```
zephyrproject\.venv\Scripts\Activate.ps1
```

Remember to activate the virtual environment every time you start a new shell instance to work on the code.

The next step is to install West:

```
pip install west
```

Once this is done, you can use it to retrieve the Zephyr source code:

```
west init zephyrproject
cd zephyrproject
west update
```

A quick command to make the Zephyr boilerplate available to CMake:

```
west zephyr-export
```

You can now fetch the SDK's reminding Python dependencies.

```
pip install -r ^
  %HOMEPATH%\zephyrproject\zephyr\scripts\requirements.txt
```

At this point, you are ready to download and install the TCK. As in the case of Linux, you need to use wget to get the SDK bundle.

```
cd %HOMEPATH%
wget https://github.com/zephyrproject-rtos/sdk-ng/releases/download/
v0.14.2/zephyr-sdk-0.14.2_windows-x86_64.zip
```

You can unpack the archive using this command:

```
unzip zephyr-sdk-0.14.2_windows-x86_64.zip
```

There are only two places where you should unpack the SDK:

- %HOMEPATH%

- %PROGRAMFILES%

You are now ready to run the SDK's setup script:

```
cd zephyr-sdk-0.14.2
setup.cmd
```

PART 3

Edge Computing and IoT Platforms

CHAPTER 11

Edge Computing

L'extrême limite de la sagesse, voilà ce que le public baptise folie.

The extreme edge of wisdom is what the public baptizes madness.

—Jean Cocteau, *L'insolence*

The "I" in IoT stands for *Internet*. And in this context, Internet means *Network*. As I wrote earlier, the focus on connectivity makes IoT devices different from embedded ones. IoT devices need to transmit the data gathered by their sensors and act on the physical world through actuators. They need a network to do that. But where is the data going, and where are the commands coming from? The Internet? Certainly not. Think about it. When you travel by car, your destination is not the road but a location you can reach by the road. When you send data to the Internet, its destination is not the Internet but the Cloud.

When I write Cloud here, I mean all its variations: public, hybrid, and private. All three offer on-demand compute, storage, and networking resources in addition to software services. What distinguishes them is their scale and location. The public Cloud is hosted in data centers owned by third-party providers; their resources and services are shared with multiple customers. Most public Cloud instances offer a scale only the largest organizations could achieve. At the other end of the scale, the private Cloud is hosted in your own facilities. What differentiates it from traditional infrastructure is that the resources are made available in a Cloud-like ("cloudish"?) way. Naturally, the scale of most private Cloud instances is much smaller than for the typical public Cloud instances. As for hybrid Cloud, it is simply a mix of the two approaches where workloads can be moved between private and public infrastructure.

© Frédéric Desbiens 2023

F. Desbiens, *Building Enterprise IoT Solutions with Eclipse IoT Technologies*,
https://doi.org/10.1007/978-1-4842-8882-5_11

Note   While the public Cloud offers bountiful resources, those are not limitless. A partner of a company I worked for in the past once exhausted all instances of a virtual machine type offered by the market-leading public Cloud provider in a specific region. We are talking about tens of thousands of virtual machines here. I am quite sure they will recognize themselves if they ever read this.

At this point, you may be asking yourself what this has to do with edge computing. You see, deploying resources and services at the Edge means that you are operating outside the confines of the Cloud and the corporate data center. In fact, from an operational standpoint, the Edge and the Cloud are opposites. But before I elaborate on this, it would be wise to define what I mean by edge computing.

What Is Edge Computing?

In its simplest form, edge computing is a form of distributed computing that brings compute, storage, and networking resources closer to the physical location where the data is produced and commands are executed. However, the simple fact of following such an architecture is not enough. You also need to build and deploy *Edge Native* applications. Such applications have several characteristics in common with Cloud Native applications:

- They rely on microservices.

- They expose APIs, often in a RESTful way.

- They are made of loosely coupled services to avoid creating affinities and to enhance the resiliency of the application.

- They are built by teams leveraging a DevOps approach, with a focus on continuous integration and continuous deployment (CI/CD).

Edge Native applications are, however, different from Cloud Native applications in a few ways. Let's see what makes the Edge and Cloud environments distinct to understand how.

Edge vs. Cloud

By nature, the Cloud is an environment that relies on an on-demand model delivered at scale. Public Clouds, in particular, offer staggering volumes of compute, storage, and networking resources. Those resources are homogeneous and correspond to predefined types. For example, all instances of a specific virtual machine type will offer the same basic capabilities, which can sometimes be expanded through configuration. While the resources can, in many cases, be geographically distributed, they are systematically administered in a centralized way.

The Edge environment is the opposite. Edge nodes are anywhere and everywhere. In other words, the Edge is the place (any place, really) where distributed computing intersects with the real world. Although edge nodes are often deployed in large quantities, the compute, storage, and networking resources available in any given location are limited. One could say that edge computing is the deployment of locally limited resources at a large scale. Edge nodes are typically heterogeneous given the operational constraints they need to face. Some must be hardened against specific environmental threats, such as water ingress, temperature, electromagnetic interference, or vibrations. Others are equipped with specialized hardware supporting specific workloads, such as artificial intelligence (AI) accelerators. Of course, most edge nodes need to keep power consumption in check because they operate on battery power or need to keep heat dissipation under control. The latter is often important since passively cooled designs are generally more reliable but require components that consume less.

The Edge-to-Cloud Continuum

If edge nodes are anywhere and everywhere, where is the edge itself, then? If you search for an answer to this question on the Internet, you will find several different ones. Table 11-1 lists a few of those answers. For each, I list the typical latency and distance between the source of the data and the node that will process it.

Table 11-1. *Commonly referenced types of edge*

Edge Type	Latency	Distance
Micro Edge	Below 1 ms	A few cm to 15 m
Deep Edge	Between 2 and 5 ms	Below 1 km
Meta Edge	Below 10 ms	Below 50 km
Fog	10–20 ms	Beyond 50 km
Multi-access Edge Computing (ETSI)	10–20 ms	100 km
Far Edge	20–50 ms	500 km

Now, you will probably ask yourself which one of those is the *real* edge. To paraphrase a famous Jedi, they are *all* real, from a certain point of view. Their main shortcoming is that they are all tied to a certain way of implementing edge computing. Thus, the Edge Native working group at the Eclipse Foundation adopted a different worldview. To them and me, edge devices are deployed along the edge-to-cloud continuum. This is also true of the applications they run.

Edge Native applications are distributed by nature. They usually encompass three distinct planes. Those planes are as follows:

- **Data plane:** Includes network, compute, and storage resources for managing data according to the commands received from the control plane.

- **Control plane:** Decides how the data plane processes and acts on data as it is received.

- **Management plane:** Configures each resource. It also monitors and maintains their state.

Please note that these definitions are not inventions on my part but are taken from RFC 7426, which is focused on the terminology of Software-Defined Networking (SDN).

The components of Edge Native applications are deployed all over the edge-to-cloud continuum independently of the plane they belong to. Figure 11-1 illustrates this reality.

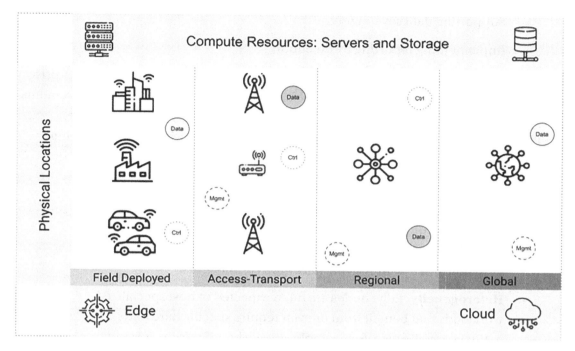

Figure 11-1. *The Edge-to-Cloud continuum (Credit: Eclipse Foundation)*

The figure highlights the first difference between Edge Native and Cloud Native applications: their relationship to physical locations. For Cloud applications, the exact physical location of the resources is irrelevant. Of course, concepts like availability zones refer to geography in a coarse-grained fashion. However, as a Cloud developer, if your goal is to deploy your code in the eastern part of Canada, you will not care if the servers are in Montréal, Québec City, or Chicoutimi. On the other hand, Edge Native applications are all about location. You deploy edge nodes to aggregate the data from *this* building, *this* factory, or *this* city. One could say that if you care about the precise location of a compute or storage resource, you are doing edge computing.

What Makes the Edge Different?

In Chapter 1, I listed a few reasons to motivate you to leverage edge computing in your projects. Those reasons were as follows:

- Optimizing bandwidth usage
- Reducing latency

275

- Supporting data sovereignty

- Implementing reporting by exception

I will not revisit them in detail here. Suffice to say that they apply to most enterprise IoT deployments, especially those that have mission-critical and real-time requirements to fulfill. Optimizing bandwidth usage is key since it is both expensive and scarce.

There are a few additional factors to consider when comparing edge computing projects to IT ones.

- **Lifespan:** Like constrained devices, edge nodes are often deployed in hard-to-reach locations. This applies whether they are under your direct control or belong to a service provider. Consequently, durability is an important consideration, as is the long-term availability of the underlying hardware components.

- **Heterogeneity:** Edge nodes are often expected to host specialized workloads that benefit from or even require specific hardware. Artificial intelligence, for example, greatly benefits from accelerators that reduce processing times and power usage. Control logic and sensor fusion involve field buses not found in run-of-the-mill IT equipment such as Modbus and Canbus. This means that the hardware configurations of edge nodes can vary a lot according to their role in the infrastructure. Consequently, you will need to consider a wide variety of processor architectures, peripherals, and networking technologies.

- **Physical constraints:** Any device you deploy in the field must face temperature, humidity, vibrations, dust, and many other threats. Naturally, this will influence the performance and durability of the hardware. For example, excessive temperatures could induce thermal throttling in the processor or other components. The software also needs to adapt to such adverse conditions. Sensor readings, for example, could need to be corrected according to current environmental factors.

- **Connectivity:** Edge computing, as a form of distributed computing, is completely at the mercy of the network. You must assume the network will degrade or fail at any time and architecture your services accordingly. Tolerance to unreliable networks is a key feature of quality edge computing implementations. This is not a given and will require careful design and deployment of the infrastructure and software components. However, the best edge computing platforms in the market will make things a bit easier.

Note My younger brother works for a large supplier of networking infrastructure whose name rhymes with Crisco. He hates it when I say we should not trust the network. However, it is a fact that the only time you notice the network is when it fails.

What Are Edge Native Applications?

At this point, we have enough perspective to take a stab at a formal definition for Edge Native applications. They are essentially microservice-based distributed applications with specific characteristics. Those characteristics include the following:

- **Optimized for field use:** Since they often run on constrained hardware, Edge Native applications are optimized for size and power consumption from the get-go.

- **Resilient:** Edge Native applications assume that individual nodes, complete services, and even the network may experience a catastrophic failure at any time.

- **Adapted to mobility:** Edge Native applications not only connect to mobile (cellular) networks but can also be deployed on nodes onboard all sorts of vehicles. This means they are not only location-aware but can leverage location-based routing when needed.

- **Orchestrated:** The components of Edge Native applications are often deployed inside containers, but virtual machines, serverless functions, and binaries can also be involved. The life cycle of all these deployment artifacts must be carefully orchestrated, whether to scale up or down certain services or to stage incremental updates involving a subset of the nodes.

- **Zero Trust security model:** Applied to Edge Native applications, the Zero Trust model implies that, by default, no device is trusted. This implies systematic device authentication and authorization while limiting the scope and timeframe of the access granted. Data protection also plays an important role in the model, meaning it should be encrypted in motion and at rest.

- **Zero Touch onboarding:** Edge Native applications require credentials for authentication, authorization, and even device attestation. The latter involves using certificates or similar means to prove a device's unique identity and trustworthiness. Manual interventions to deploy credentials on devices pose a significant security risk. Zero Touch onboarding means that such credentials can be deployed from a central location as soon as a device connects to the network. This tremendously simplifies the process of deploying infrastructure at a large scale since human interventions are reduced to a minimum. Deployment costs are also significantly lower, given that the process is streamlined and automated.

Many of the characteristics I just presented involve a certain level of complexity. Edge computing platforms can relieve developers from part of the burden since they can rely on platform features rather than implementing the logic themselves. Moreover, since they offer generic implementations of common features, the platforms can drive the adoption of specific software development patterns for developers of all skill levels across industries. Edge computing platforms allow you to learn software development for the Edge in a structured way.

The Need for EdgeOps

DevOps transformed the way we build, deploy, and operate applications. Like Agile, one of its inspirations, DevOps is an umbrella term for a loosely related set of practices that most organizations tweak and shape according to their culture and priorities. The main goal of DevOps is to shorten the development life cycle through continuous delivery. To achieve this, it strives to break the traditional division of labor between developers and system administrators (operators). In some organizations, this was implemented by merging teams of developers and administrators; in others, distinctions between functions were maintained, but barriers to collaboration were eliminated.

At the technical level, DevOps favors microservices deployed in containers; those containers are built by continuous integration pipelines and, through continuous deployment, made available to end users almost immediately. The goal is to avoid the past operational models, which were characterized by the infrequent deployment of large releases of monolithic software components. DevOps, on the other hand, emphasizes the incremental delivery of small changes in real time. If you use any modern software-as-a-service (SaaS) web platform, you probably were invited to reload the page to benefit from new features at some point. This is a visible hint of DevOps in action behind the scenes.

EdgeOps also relies on container-based microservices as well as continuous integration. However, differences appear in the deployment model. Continuous deployment is not desirable or even possible for many edge computing use cases, such as connected vehicles, industrial automation, and patient monitoring. Generally, use cases involving mission-critical and real-time requirements rely on cautious, incremental deployment models for software updates. In addition, since there is little to no elasticity at the Edge, new releases are carefully tested for performance and power consumption before being pushed to production nodes. Direct over-the-air (OTA) updates are sometimes impossible since the nodes operate in "air-gapped" environments that are not connected directly to the public Internet or the corporate network. In such cases, deploying a new version of a software component requires dispatching personnel on-site; nodes will then be updated one by one or through an OTA service present in the local environment. Suffice to say that update velocity is typically much lower at the Edge. In that regard, it is much closer to operational technology than information technology.

EdgeOps is an evolution of DevOps that considers the IT challenges edge computing addresses and the specific characteristics of edge computing solutions. In other words, it is DevOps but tweaked to support Edge Native applications. Like DevOps, it is a set of practices; however, such practices are reflected in the feature set of best-of-breed edge computing platforms.

Edge Computing Platforms

Nowadays, Kubernetes seems to be the answer whatever the question is. This is true at the Edge as it is in the Cloud. However, running plain Kubernetes (K8s) or its commercial distributions at the Edge only makes sense when sufficient compute resources are available. On constrained nodes, it is preferable to deploy lighter-weight versions such as K3s, KubeEdge, or MicroShift. It would be shortsighted to restrict your search for an edge computing platform to K8s and its derivatives. Some workloads do not lend themselves easily to containers since they require direct interaction with the hardware. Also, there are specific benefits to platforms built from the ground up for the Edge, such as the capacity to run in environments not connected to the Cloud and the automatic selection of container images fitting the hardware profile of a particular node. For these reasons and many others, you should evaluate a broad set of options when selecting an edge computing platform.

Table 11-2 compares the main characteristics of several edge computing platforms.

Table 11-2. *Characteristics of popular edge computing platforms[1]*

Platform	Cloud Managed	Edge Only	K8s Integration	Focus
AWS Outposts	Yes	No	Offers K8s	Containers, VMs
Eclipse ioFog	Yes	Yes	Yes	Containers
Eclipse Kanto	Yes	Yes	No	Containers, IoT
Eclipse Kura	Yes	Yes	No	Containers, Gateways
EdgeX Foundry	Yes	No	No	IoT
Fledge	Yes	No	No	Industry 4.0
K3s	No	Yes	Is K8s	Containers
KubeEdge	Yes	Possible	Is K8s	Containers
Open Horizon	Yes	No	Yes	Containers

As you can see, a wealth of platforms are available, each with its focus and feature set. Three of them call the Eclipse Foundation their home: Eclipse ioFog, Eclipse Kanto, and Eclipse Kura. I will now explain in detail their architecture and feature set. Of course, I will also provide an overview of the main edge-focused distributions of Kubernetes. Finally, I will introduce you to project Eve, an intriguing approach to operating systems for edge nodes.

---

[1] Adapted from G. Baldoni, L. Cominardi, M. Groshev, A. De la Oliva, and A. Corsaro, "Managing the far-Edge: are today's centralized solutions a good fit?"

Eclipse ioFog

Eclipse ioFog is a mature edge computing platform for deploying, running, and networking distributed microservices at the Edge. It has been contributed to the Eclipse Foundation by Edgeworx, a startup based in California.

The official web resources for ioFog are as follows:

- **Website:** `https://iofog.org`

- **Eclipse project page:** `https://projects.eclipse.org/projects/iot.iofog`

- **Code repositories:** `https://github.com/eclipse-iofog`

At the time of writing, the team was about to release ioFog version 3, which had been in beta for several months. This section is based on version 3 since it significantly improves the platform. ioFog is made available under the Eclipse Public License version 2.

ioFog Concepts and Architecture

The core concept of ioFog is the Edge Compute Network (ECN). Each ECN is made of one or more nodes. Nodes are devices running Linux on which the ioFog Agent is deployed. The Agent is a daemon responsible for the set of microservices running on a specific node. Another ioFog component, the Controller, is responsible for orchestrating the agents. It represents ioFog's control plane.

By default, the Controller is deployed on a stand-alone host. This is what the ioFog documentation refers to as a *remote* deployment. It is also possible to deploy the controller onto a Kubernetes cluster. In that case, ioFog relies on a custom Kubernetes operator[2] to integrate with the cluster and deploys resources for routing. Please note that the ioFog control plane will work on most Kubernetes distributions, including managed ones. However, lightweight implementations such as MicroK8s and K3s are not supported – although Minikube is. In ioFog, microservices are always deployed inside containers. The life cycle of those containers is under the control of the ioFog Agent, which relies on the Docker container runtime.

Figure 11-2 illustrates how those concepts all fit together in the ioFog platform when the Controller is deployed on a stand-alone server (remote deployment).

---

[2] `https://github.com/eclipse-iofog/iofog-operator`

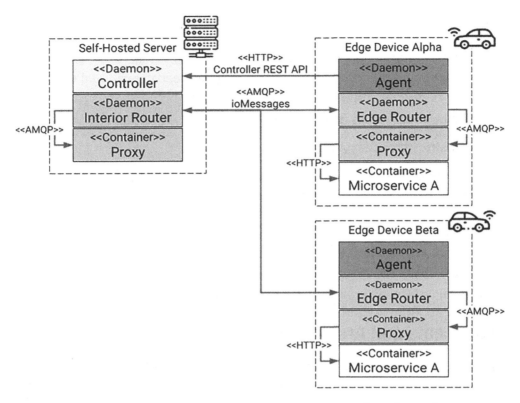

Figure 11-2. *ioFog architecture for remote deployments (Credit: Eclipse Foundation)*

Another important ioFog concept is *Applications*. Applications represent a set of microservices deployed together for a particular purpose. Both applications and microservices are described using YAML definition files. Applications play an important role because, by default, microservices cannot communicate with one another. An application specifies its name, microservice instances, and the routes between the microservices. Without a route, network traffic will be blocked. ioFog version 3 also introduced powerful templating mechanisms for applications.

The ioFog Controller provides a complete REST API and a graphical web interface called ECN Viewer to inspect the state of every agent, application, and microservice. You can see a screenshot of ECN Viewer in Figure 11-3.

Figure 11-3. *Screenshot of ioFog's ECN Viewer*

When working with ioFog, you will spend most of your time working with the `iofogctl` command-line tool. You can use it to create, configure, and operate ECNs and their resources, such as Controllers, Agents, Applications, and microservices.

Getting Started with ioFog

ioFog itself runs on Linux. However, there are versions of `iofogctl` for Linux, macOS, and Windows. In the latter case, all you need to do is to download the executable and place it in your path.

To install ioFog on macOS using the Homebrew package manager, issue the following commands:

```
brew tap eclipse-iofog/iofogctl
brew install iofogctl@3.0
```

On Linux, the ioFog team maintains package repositories for Debian and RPM-based distributions. You also have the choice to install the binary directly. To do that, execute these commands:

```
curl -LO \
  https://storage.googleapis.com/iofogctl/linux/3.0/iofogctl
sudo install -o root -g root -m 0755 iofogctl \
  /usr/local/bin/iofogctl
rm ./iofogctl
```

You can test that the CLI is installed properly by running `iofogctl version`. The output should look like this:

```
version: 3.0.1
platform: linux/amd64
commit: 302077d2d66a0d788e7b6de99e65be8dc633f09e
date: 2022-05-27T02:06:35+0000
```

To deploy the controller and an agent instance on your workstation for testing purposes, paste the following YAML into a file in a suitable location. The ioFog documentation suggests `/tmp/quickstart.yaml`.

```
---
apiVersion: iofog.org/v3
kind: LocalControlPlane
metadata:
  name: ecn
spec:
  iofogUser:
    name: Quick
    surname: Start
    email: user@domain.com
    password: q1u45ic9kst563art
  controller:
    container:
      image: iofog/controller:3.0.0
---
apiVersion: iofog.org/v3
```

```
kind: LocalAgent
metadata:
  name: local-agent
spec:
  container:
    image: iofog/agent:3.0.0
```

Then, you can deploy the controller and agent by executing this command:

```
iofogctl deploy -f /tmp/quick-start.yaml
```

Once the containers are up, you can access ECN Viewer at the following URL:

```
http://localhost:8008/#/overview
```

The username and password were specified in the preceding YAML descriptor.

Eclipse Kanto

Eclipse Kanto is the newest edge computing platform contributed to the Eclipse Foundation. The project was initiated by Bosch.IO, a subsidiary of Bosch focused on the Internet of Things. Because of this, it sports stellar integration with several other Eclipse IoT projects, such as Eclipse hawkBit (software updates), Eclipse Hono (device connectivity), and Eclipse Ditto (digital twins). Kanto is available under the Eclipse Public License version 2. At the time of writing, only the initial release for Kanto, v. 0.1.0-M1, was available.

The official web resources for Kanto are as follows:

- **Website:** https://eclipse.org/kanto

- **Eclipse project page:** https://projects.eclipse.org/projects/iot.kanto

- **Code repositories:** https://github.com/eclipse-kanto

Kanto Concepts and Architecture

The project team describes Kanto as a modular platform delivering essential features to IoT devices like Cloud connectivity, digital twins, local communication, container management, and software updates. The devices are remotely configurable and manageable from a choice of Cloud IoT platforms. Figure 11-4 summarizes Kanto's architecture. The platform heavily relies on the MQTT protocol and bundles the Eclipse Mosquitto broker to handle internal message passing between the components.

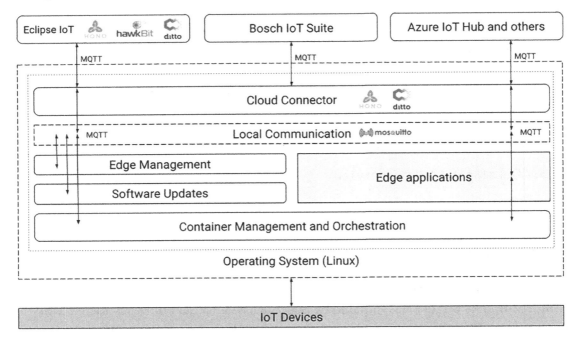

Figure 11-4. *Eclipse Kanto architecture (Credit: Eclipse Foundation)*

As you can see, the platform has four major functional blocks. Those blocks are as follows:

- **Connector:** Handles Cloud connectivity. It can interact with Eclipse IoT open source IoT platforms (Ditto, hawkBit, Hono), their commercial equivalents in the Bosch IoT Suite, or Microsoft's Azure IoT Hub.

- **Container management:** Exposes a unified API abstracting the core operations of container management, including life cycle, state, networking, host resources access, and usage management. Kanto supports multiple container runtimes, whether OCI compliant (containerd, runc, kata) or Linux native (LXC, LXD).

- **File uploads:** Provides file uploads to AWS s3 buckets or plain HTTP servers. Kanto can upload the files at specified intervals or wait for an explicit trigger sent by the Cloud back end. The feature supports a variety of use cases such as diagnostics, monitoring, node backup, and system restore.

- **Software updates:** Downloads, validates, and installs software updates for edge nodes and devices connected to them. Kanto can monitor the progress of those operations and can resume interrupted downloads.

System administrators often interact with Eclipse Kanto through another IoT platform. The project's website provides examples for Eclipse Hono and the commercial Bosch IoT Suite.

Getting Started with Kanto

For the time being, you can deploy Kanto on Linux only. Binaries are published for the x86_64, armv7 (32-bit), and arm64 architectures as Debian packages (.deb). You can access the releases on the project's main GitHub repository: `https://github.com/eclipse-kanto/kanto/releases`.

The only prerequisite for Kanto at this point is the `containerd` runtime. You can install it using your distribution's package manager or the script provided by the Kanto team, as shown here:

```
curl -fsSL https://github.com/eclipse-kanto/kanto/raw/main/quickstart/
install_ctrd.sh | sh
```

To install Kanto itself, download the .deb matching the architecture of your node and then install it using apt. For example:

```
sudo apt install ./kanto_0.1.0-M1_linux_x86_64.deb
```

Each of the components will be installed as a service. You can check their status with this command:

```
systemctl status \
suite-connector.service \
container-management.service \
software-update.service \
file-upload.service
```

287

Kanto is now ready to use! Using the public Eclipse Hono sandbox is a good way to exercise it. The Kanto team provides instructions to that effect here: `www.eclipse.org/kanto/docs/getting-started/hono/`. Naturally, you could also deploy a local instance of Hono using the Cloud2Edge package from the Eclipse IoT Packages project.

Eclipse Kura

Are IoT gateways edge nodes? In many ways, yes, although traditional gateways could not run containerized Edge Native applications. This changed in 2021 when Eclipse Kura gained a container orchestration provider.

One of the very first projects of Eclipse IoT, Eclipse Kura, has been going strong since 2013. Created initially by Eurotech, the project attracted several third-party contributions throughout the years. It is available under the Eclipse Public License version 2. When I wrote this, the latest release was version 5.1.1, published on April 26, 2022, and candidate builds were available for version 5.1.2. Eclipse Kura is at the core of Everyware Software Framework (ESF), a commercial product from Eurotech.

The official web resources for Kura are as follows:

- **Website:** `https://eclipse.org/kura`

- **Eclipse project page:** `https://projects.eclipse.org/projects/iot.kura`

- **Code repositories:** `https://github.com/eclipse-kura`

Kura Concepts and Architecture

Eclipse Kura is a platform and application framework for building smart connected Edge systems and IoT gateways. It offers a remote management interface and exposes many APIs supporting IoT applications. Kura is completed by Eclipse Kapua, an IoT platform providing a data and device management back end for Kura-powered edge devices. I will cover Kapua in the next chapter. Kura is written in Java; all it needs to run is a Java Standard edition runtime on a Linux operating system. The platform leverages the OSGi component model, which streamlines the process of writing reusable software building blocks. Kura sports deep integration with hardware, including serial ports, GPS, USB, GPIO, I²C, and others.

Figure 11-5 illustrates Kura's architecture and feature set.

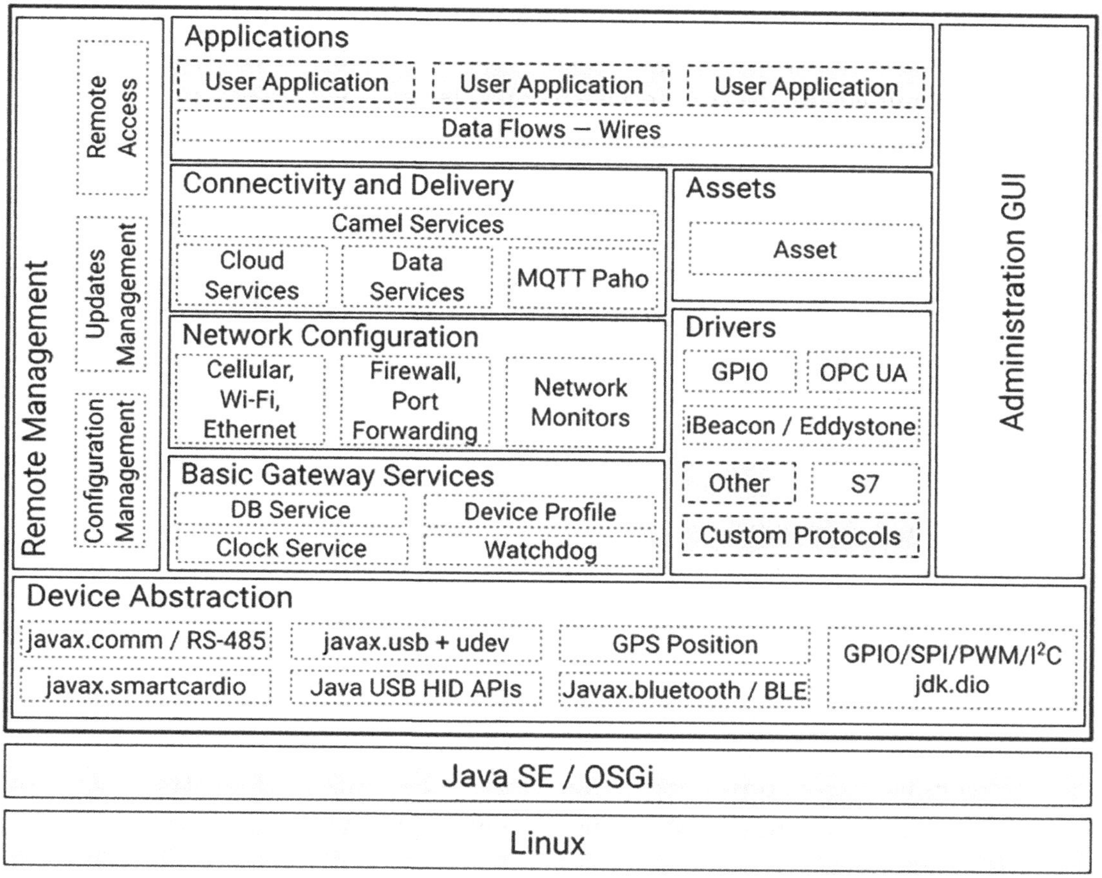

Figure 11-5. *Eclipse Kura architecture (Credit: Eclipse Foundation)*

Kura offers developers an extensive set of services, including configuration, device life cycle management, remote access features, log management, health monitoring, networking, and Edge AI enablement. Here are additional details about a few other standout services:

- **Data:** Kura defines generic data APIs abstracting the underlying database. This provides the flexibility to change the SQL database provider without changing the applications themselves. Kura offers a built-in SQL database engine (H2) and MQTT Broker (ActiveMQ Artemis). You can also deploy Apache Camel locally or integrate with a remote instance. Camel provides a rules-based routing and mediation engine. Kura also provides transient storage for messages.

This allows it to cope with network disruptions in distributed edge deployments. Messages persisted locally that could not be transmitted are preserved even when the device reboots. They are sent to the Cloud once network conditions allow it. This ensures that the telemetry data history is preserved.

- **Cloud:** Kura's Cloud APIs abstract the actual Cloud provider; applications are not aware of the actual provider used. This makes it easier to switch providers if needed. Connections to remote Cloud services can be shared among several applications deployed on the gateway. Connecting to multiple Clouds at once (multi-cloud) is also possible. Out of the box, Kura can connect to the following Cloud services: AWS IoT, Azure IoT Hub, and Eurotech Everyware Cloud platform. Additional connectors are available through the Eclipse Marketplace, and you can also develop custom ones. Kura-based gateways are ready to connect to local or remote instances of the Eclipse Kapua IoT platform and the Eclipse Hono device connectivity platform.

- **Security:** Kura is developed with security in mind. Eurotech's commercial offering based on Kura, combined with one of its latest IoT Gateway hardware platforms, has obtained PSA Level 1 and IEC 62443-4-2 certifications.

- **Wires:** This innovative feature provides a visual data flow programming tool to define data collection and processing pipelines.

From a development standpoint, Kura applications are written in Java and packaged in OSGi bundles. However, the recently added support for containers means you can also build applications with other languages. Such containerized applications can interact with the Kura framework using REST APIs and JDBC or connect directly to the AMQ broker and Camel if deployed.

Kura also ships with a rich web graphical interface. Figure 11-6 showcases the Wire Graph editor.

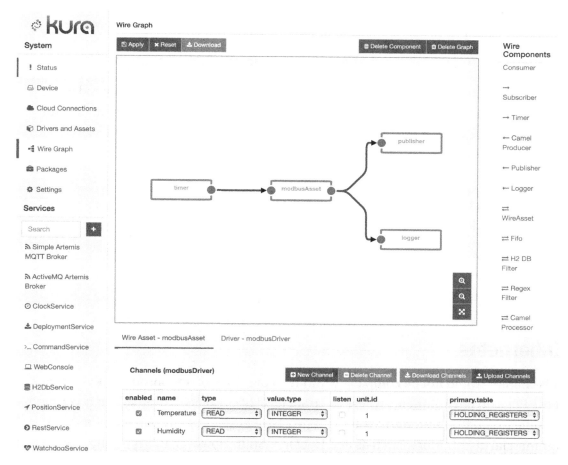

Figure 11-6. *Eclipse Kura graphical interface (Credit: Eurotech)*

Getting Started with Kura

The fastest way to deploy a Kura instance is to use the container image maintained by the project team and published to Docker Hub. On a workstation where Docker is already installed, all you need to do is to execute the following command:

```
docker run -d -p 443:443 -t eclipse/kura
```

Once the container is up, you can access the web user interface at `https://localhost`. The default credentials are `admin/admin`.

If you prefer to test drive on a Raspberry Pi, the team provides `.deb` packages for Raspbian and Ubuntu. You can download them from the following location: `http://download.eclipse.org/kura/releases/`.

Download the latest Kura release matching your OS on a freshly updated system. You would, for example, install the Raspbian version using this command:

```
sudo apt-get install ./kura_<version>_raspberry-pi_installer.deb
```

On Raspbian, you need to ensure the SSH daemon is started. You also need to disable consistent device naming by adding the `net.ifnames=0` parameter at the end of the `/boot/cmdline.txt` file.

The Kura documentation provides additional guidance and caveats. You will find full details here:

- **Raspbian:** `http://eclipse.github.io/kura/intro/raspberry-pi-quick-start.html`

- **Ubuntu:** `http://eclipse.github.io/kura/intro/raspberry-pi-ubuntu-20-quick-start.html`

The team also produces a build for Nvidia's Jetson Nano platform.

Kubernetes

This chapter would not be complete without an overview of options to run Kubernetes at the Edge. In environments where a certain level of elasticity is available, it is certainly possible to run the open source distribution of Kubernetes or any of the available commercial distributions, such as Red Hat OpenShift. However, this typically requires deploying servers in a variety of data center. My focus in this section will be on lighter-weight alternatives that can run on nodes deployed *in the wild*, specifically K3s and KubeEdge. There are, of course, several others, such as Canonical's MicroK8s or Red Hat MicroShift.

Note   If you are less familiar with the platform, Kubernetes is a container orchestration system automating software deployment, scaling, and management. It was originally built by Google and is now maintained by the Cloud Native Computing Foundation (CNCF), which belongs to the Linux Foundation. As of 2022, Kubernetes is a *de facto* industry standard to orchestrate Cloud Native applications.

K3s and KubeEdge represent two distinct ways of slimming down Kubernetes for Edge environments. Their governance model is also different. The former is a project initiated by Rancher Labs (since then purchased by SUSE). The latter is under the stewardship of the CNCF. Both code bases are made available under the Apache License version 2.

K3s packages every component as a single binary and can leverage the SQLite embedded database instead of etcd, which reduces the runtime footprint. It also removes a few features from upstream Kubernetes, namely, the cloud provider and storage drivers. K3s diverges slightly from the upstream code base, but the changes are intentionally minimal. The project explicitly intends not to change any core Kubernetes functionality.

KubeEdge takes a different approach. It provides a set of Cloud and Edge components working together to extend Kubernetes at the Edge. KubeEdge permits the deployment and orchestration of services on edge nodes and metadata synchronization with the Cloud. It also enables the centralized management of edge nodes from a Cloud-based Kubernetes control plane. The runtime footprint of the Edged agent, which manages containerized applications on the edge nodes, can be as low as 10 MiB.

Figure 11-7 contrasts the architectures of K3s and KubeEdge with the upstream Kubernetes distribution.

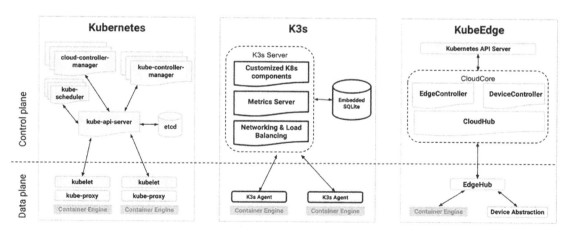

Figure 11-7. *Comparison of the Kubernetes, K3s, and KubeEdge architectures*[3]

---

[3] Adapted from G. Baldoni, L. Cominardi, M. Groshev, A. De la Oliva, and A. Corsaro, "Managing the far-Edge: are today's centralized solutions a good fit?"

K3s, KubeEdge, and their alternatives represent effective ways to orchestrate containers running on edge nodes. However, such platforms induce additional latency, making them a poor fit for use cases involving mission-critical and real-time requirements. Kubernetes, after all, works best with stateless workloads and microservices. Many Edge applications are stateful by design.

Project EVE

All the platforms I covered up to now run on the top of an operating system, typically Linux. Project EVE, from LF Edge, goes further since it integrates application orchestration to the base operating system. EVE and its companion projects were contributed to LF Edge by Zededa: a California startup focused on edge computing solutions. The code base is available on GitHub under the Apache License version 2.

The official web resources for EVE are as follows:

- **Website:** www.lfedge.org/projects/eve/

- **Code repository:** https://github.com/lf-edge/eve

EVE stands for *Edge Virtualization Engine* and supports virtual machines and containers through a level 1 hypervisor (currently based on Xen) running on a low-footprint and hardened Linux kernel. Since it relies on a hypervisor, EVE is more suitable for use cases involving mission-critical and real-time requirements than other solutions leveraging KVM, the Linux kernel's built-in virtualization solution. Moreover, EVE features deep integration with hardware root of trust solutions, providing a strong foundation for device attestation and other security measures when running directly on physical hardware (bare metal). It also can disable unused ports on edge nodes to prevent device tampering.

The EVE runtime does not possess a user interface to keep it lightweight. Edge nodes are managed through a RESTful API. On the other hand, management functions are bundled into a central controller. The EVE team maintains an open source controller implementation known as Adam. It implements essential features and is a good way to start with the platform. This separation of concerns between the node runtime and controller means that users can pick between open source and commercial controllers, such as Zededa's own, depending on their needs. For security reasons, EVE nodes are associated with a specific controller when they are initially provisioned, and this association cannot be changed.

Getting Started with Eve

The easiest way to get started with EVE is to rely on a companion project called Eden. Eden makes it easy to deploy an instance of the Adam controller and a set of EVE nodes. The minimum configuration requires two devices, which can be actual hardware or virtualized. One will run Eden and the other Eve. In this context, the device hosting the Eden components is called the *manager*. The basic prerequisites you need on the manager are the Eden binary, which you can download from its GitHub repository, and a functional install of the Docker runtime. On Linux, you need to run the following commands to ensure Docker will have the ability to execute commands:

```
sudo usermod -aG docker $USER
newgrp docker
```

The Eden manager runs several important components. Those components are as follows:

- **Adam:** The controller, running as a daemon

- **Redis:** A database instance used for logging, running as a daemon

- **Eserver:** A repository of files and EVE device images made available for download under the HTTP and FTP protocols

- **Registry:** An OCI-compliant container registry to make images available to edge nodes

The device running EVE has a different set of prerequisites. Here is a comprehensive list for a virtual one – meaning that you will run the manager and EVE node in parallel on the same physical hardware:

- qemu, version 4.x or higher

- telnet

- squashfs tools (squashfs or squashfs-tools, depending on the OS used)

On Linux, you also need to make sure KVM is available and can execute commands. After installing it (if needed), execute these commands:

```
sudo usermod -aG kvm $USER
newgrp kvm
```

On macOS, you will need either machyve or Parallels to provide virtualization.

I will now provide instructions on setting up the infrastructure and deploying an EVE image running the Nginx web server. The first step is to create a default configuration:

```
eden config add default
```

Then, execute these commands to set up the manager's components and start them:

```
eden setup
eden start
```

Since we now have a running controller and one EVE node, it is now possible to associate the two:

```
eden eve onboard
```

The next step is to deploy an Nginx container on EVE. Notice that we mount a local folder to provide content to serve to the web server. This supposes you cloned the Eden repository and are running the command from the top-level folder of the clone.

```
eden pod deploy docker://nginx -p 8028:80 --mount=src=./data/helloeve,dst=/
usr/share/nginx/html
```

Finally, you can check that everything is running properly with the following command:

```
eden status
```

Time to access the web content served from the EVE node. This command should do the trick:

```
curl http://<EVE IP>:8028
```

You now have a working Eden environment that you can use to explore the possibilities of EVE. If you wish to stop the components and clean up, execute this command:

```
eden stop && eden clean --current-context=false
```

CHAPTER 12

Applications

Il y a très loin de la velléité à la volonté, de la volonté à la résolution, de la résolution au choix des moyens, du choix des moyens à l'application.

There is a difference between feebleness by the impotence of the will, of the will to the resolution, of the resolution to the choice of means, of the choice of the means to the application.

—Jean François Paul de Gondi, cardinal de Retz

In the 1990s, when I still had (some) hair on my head, the IT landscape was much more diverse. Several flavors of UNIX fought for dominance. Multiple processor architectures that are now gone or on their way to obsolescence, such as DEC's Alpha, HP's PA-RISC, and Sun's SPARC, were common sights in data centers. On the desktop, IBM's OS/2 was the OS of the future until Windows stole its lunch. As I write this in 2022, only a handful of viable options remain in each category. Why is that? Which factor ultimately determined the surviving processor architectures and operating systems? Naturally, the business strategies of the organizations involved played a role, as did the quality and features of the technologies involved. However, the decisive element was something else. In the end, users picked the processors and operating systems that enabled them to run the applications they wanted. In that sense, Word and Excel did more to ensure Microsoft's dominance on the desktop than Windows itself.

Microcontrollers, edge nodes, protocols, and operating systems: these building blocks are not ends of their own but rather ways to run applications. This chapter will describe three specific open source Edge applications or application frameworks hosted at the Eclipse Foundation. The first one, Eclipse 4diac, is a design environment and runtime for programmable logic controllers. The second one is Eclipse Keyple, a technology enabling contactless payments. Finally, I will cover Eclipse VOLTTRON, a distributed sensing platform with an energy management slant.

© Frédéric Desbiens 2023
F. Desbiens, *Building Enterprise IoT Solutions with Eclipse IoT Technologies*,
https://doi.org/10.1007/978-1-4842-8882-5_12

If you are a developer, you probably are or will be building such applications yourself. Or maybe you will work on an IoT platform of your own. If that is the case, you must pick a runtime to leverage in your Edge and Cloud applications. I will discuss the most widely used.

Whatever your choice of runtime is or will be, you will need to integrate devices using a variety of protocols into the rest of your infrastructure. There are two main ways to achieve this. The first is using a device connectivity platform such as Eclipse Hono. Such platforms provide uniform interfaces that business applications can use to receive telemetry and send commands. The platform then abstracts the IoT protocols used. The second approach is to have the devices themselves expose a protocol-agnostic interface. A good example is the Web of Things (WoT) specification from the World Wide Web Consortium (W3C). Both approaches support the digital twins pattern, which is probably something you will want to leverage as well.

In this chapter, I will focus on the first approach. I will discuss WoT, digital twins, and the relevant Eclipse IoT projects. In the next chapter, I will cover IoT platforms, including device connectivity ones.

Application Runtimes

Let me be clear: all the runtimes I mention in this section are good options for writing an Edge application or Cloud IoT platform. That said, some of them may not be optimal choices for targeting constrained edge nodes. Moreover, the robustness of their open source ecosystem for IoT protocol implementations differs in terms of variety, quality, and maturity. I will try to cover those aspects for each of them.

.NET

Traditionally, Microsoft's development tools and runtimes were proprietary and only supported the Windows operating system. This changed in 2016 with the introduction of .NET Core, the open source version of the framework. Microsoft even established the .NET Foundation to provide vendor-neutral governance around it. This undoubtedly spurred an increased interest in it from open source contributors. In 2020, Microsoft announced that .NET Core would be rebranded as simply .NET and that all future work would go in that renamed, open source version. As of 2022, it is possible to write and deploy code targeting .NET on Linux, macOS, and Windows. The runtime supports the C#, F#, and Visual Basic programming languages.

Note   You probably are still confused about .NET's rebranding. Here is an easy way to distinguish the various versions. In Microsoft's materials, the proprietary Windows-only version of .NET is called .NET Framework. The current version is 4.8. The open source multiplatform version is called .NET, formerly .NET Core. Version 5.x and up of .NET refer to the open source version.

From a purely technical perspective, .NET is a fine framework, and the C# and F# language implementations are top-notch. However, the availability of open source IoT protocol implementations is spotty. For example, the C# implementation of the Eclipse Paho MQTT client lacks regular contributors and does not support MQTT v5. Naturally, the size of the full .NET Core runtime is also a problem on constrained devices. Also, although .NET Core is open source and under vendor-neutral governance, Microsoft still plays an outsized role in the overall ecosystem.

Java SE and Jakarta EE

The Java virtual machine (Java Standard Edition) and the Java Enterprise Edition (Java EE) runtime entered the market as proprietary products from Sun Microsystems. In 2007, Sun made the Java Standard Edition runtime and development kit open source, giving birth to OpenJDK. In 2018, Oracle decided to open-source the Java Enterprise Edition specifications at the Eclipse Foundation. The technology was renamed Jakarta EE since Oracle decided to keep ownership of the original Java Enterprise Edition trademark.

Note   If you are looking to download prebuilt OpenJDK binaries, you should have a look at the Adoptium Marketplace. The Marketplace promotes high-quality TCK-certified and AQAvit-verified runtimes from various vendors supporting several architectures. You can find it at: `https://adoptium.net/marketplace`. Adoptium is a working group of the Eclipse Foundation, and its own OpenJDK distribution is called Eclipse Temurin.

Although Java is not as trendy as it was around 1998, its ecosystem offers a wealth of high-quality and mature IoT protocol implementations. Many of the open source IoT platforms found at the Eclipse Foundation are built with it. Moreover, the Java virtual machine supports languages other than Java, such as Kotlin, Ruby, and Scala. It is possible to use them while leveraging the existing ecosystem of Java libraries.

One persistent criticism against Java has been the startup times of the virtual machine, which made the runtime less than ideal for implementing stateless microservices deployed to containers. This issue was compounded by the sheer size of the Java EE/Jakarta EE runtime. Several efforts were made over the years to make Java SE more modular and slim down Jakarta EE. A good example of the latter is Microprofile, which is an open source specification for Enterprise Java microservices. Microprofile runtime implementations only include a strict subset of the full Jakarta EE platform in support of JSON REST microservices. Other open source projects, such as Red Hat's Quarkus, perform as many operations as possible at build time to include only the classes actually used at runtime in the build output. Quarkus also emphasizes compilation to a native executable through GraalVM or its Mandrel downstream distribution.

Even though its syntax is more verbose than younger programming languages, Java is well worth your attention, given its broad support and maturity. Innovations such as Microprofile and Quarkus address several of its weaknesses. The release cadence for new language versions is much more predictable nowadays, and development is performed in the open. There are several implementations of Java SE and Jakarta EE, many of which are open source.

Node.js

JavaScript is one of the core technologies of the World Wide Web. Like HTML and CSS, its roots go back to the 1990s. Brendan Eich of Netscape originally designed the language. Nearly all modern websites use JavaScript, and all major web browsers possess a dedicated JavaScript engine to execute client-side logic. JavaScript conforms to the ECMAScript standard, which defines the core features of a scripting language. ECMAScript does not cover I/O facilities such as networking, storage, or graphics. Most JavaScript engines provide such features on their own.

> **Note**   JavaScript's syntax resembles Java's, but the two languages never had a direct relationship. The original name picked by Netscape was LiveScript, which was changed in December 1995 to capitalize on Java's market momentum. To add to the confusion, Sun Microsystems trademarked the name JavaScript, and Oracle still owns that trademark.

In 2009, JavaScript on the server became a serious option with the inception of the Node.js runtime environment. Node.js is an open source runtime built upon an event-driven architecture enabling asynchronous I/O. Node.js applications are interpreted. The runtime is available on most modern operating systems, including Linux, macOS, and Windows. Node.js only supports JavaScript, but several languages transpile to it, such as CoffeeScript, Dart, TypeScript, and many others. Node.js is under the vendor-neutral governance of the OpenJS Foundation, which is affiliated with the Linux Foundation.

> **Note**   A transpiler is a software that takes source code written in a language and produces the equivalent source code in a different language. So, for example, although JavaScript is loosely typed, TypeScript has a strong type system.

Node.js shines for applications that require nonblocking I/O and asynchronous request handling. From that perspective, it is a good fit for IoT applications and microservices. Older versions had trouble dealing with processor-bound tasks, but the introduction of worker threads in version 12 mitigated that issue. The fact that the code is interpreted does not mean that performance is lower or the runtime footprint is higher. Although the ecosystem of modules (packages) for Node.js is very large, many do not benefit from timely security updates or regular maintenance. In other words, because a module is available in the index of Node's package manager (NPM) does not mean you should blindly trust it. Moreover, intellectual property issues involving code provenance or incompatible licenses are not infrequent. If you work with Node.js, you will need to give extra scrutiny to your direct and indirect dependencies.

There are JavaScript implementations for the most popular IoT protocols. When selecting one, make sure that it is compatible with the specific version of Node.js you intend to use. You should also thoroughly examine all direct and indirect dependencies as possible to ensure you can trust their origin and use licenses compatible with each other. Due to our strong IP management processes, JavaScript modules offered by the Eclipse Foundation do not typically suffer from such issues.

Python

Python is both a programming language and a runtime. It is the brainchild of Guido van Rossum, who started working on Python in the late 1980s and published its first release in 1991. Python is under the stewardship of the Python Software Foundation, created in 2001.

Like Java and .NET, Python is compiled to bytecode. The reference implementation of the runtime, CPython, executes that bytecode on its virtual machine. Several alternate production-quality runtimes exist for Python: Jython, written in Java for the Java virtual machine (JVM); PyPy, written in RPython and translated into C; and IronPython, which is written in C#. Those alternate runtimes support the syntax and features of the Python version they emulate but may miss features or exhibit differences in behavior. Moreover, the MicroPython and CircuitPython variants possess runtimes optimized to run on microcontrollers. Be aware, however, that those alternate runtimes do not necessarily implement all the language features found in CPython; this impacts the compatibility of the libraries found in the ecosystem with a specific runtime.

This wide variety of runtimes means that Python is one of the most widely available languages in the industry. CPython, for example, supports not only Linux, macOS, and Windows but also several less common operating systems such as AIX, FreeBSD, Solaris, and z/OS. The runtime variants targeted at resource-constrained environments extend the reach of Python further and even provide low-level access to hardware. However, Python is generally seen as unsuitable for computationally intensive applications. This, of course, depends on the actual runtime selected and how it is configured.

Python offers a comprehensive standard library and an enormous ecosystem of third-party packages. It is a *de facto* standard in data science, and there are well-maintained implementations of all the IoT protocols mentioned in this book. The Python version of Eclipse Paho is even in the top 1% of most downloaded packages. Consequently, Python is well worth your attention as a runtime for IoT applications. However, the Python Package Index (PyPI) suffers from the same issues as NPM in the Node.js ecosystem: packages of dubious origins or lacking maintenance abound, and licensing or other IP issues are also frequent. In July 2022, the Python Software Foundation announced that the maintainers of packages identified as critical would need to use two-factor authentication (2FA) to prevent the hijacking of their code by malicious actors.[1]

Rust

Rust, one of the most recent entries in this section, is a compiled programming language. It started in 2006 as a personal project of Graydon Hoare, an employee of Mozilla. The first numbered release of the compiler was released in 2012. Rust focuses on type safety and concurrency. Rust provides memory safety; all references point to valid memory at runtime. This is enforced at compilation time. Rust is owned and maintained by the Rust Foundation, created in 2021.

Contrarily to the other languages mentioned in this section, Rust does not offer a runtime in the conventional sense. Parts of the Rust standard library, which is comprehensive, provide features usually associated with a runtime, such as a heap, backtraces, unwinding, and stack guards. The standard library (or crate, in Rust parlance) assumes the application will be deployed on a desktop-class or server-class operating system such as Linux. In addition, the Rust standard library also links to the C standard library. For applications deployed on bare metal microcontrollers, it is also possible to eschew the standard crate and link to the core crate instead. The core crate is a platform-agnostic subset of the standard crate and makes no assumptions about the system the program will execute on. Thus, Rust programs relying only on the core crate can be lower-level components such as bootloaders, firmware, or operating system kernels.

Rust was included in this section due to its growing popularity in IoT and edge computing circles. Projects such as Drogue IoT show Rust's suitability to write applications for constrained devices, the Edge, or the Cloud. In Rust, packages of code are referred to as *crates*. Since Python and JavaScript have been around for much longer, their respective library ecosystem dwarfs Rust. However, there are well-maintained implementations for most IoT protocols. As a reminder, the core implementation of the Eclipse Foundation's own zenoh protocol is written in Rust.

---

[1] See https://pypi.org/security-key-giveaway/ for more details.

WebAssembly

WebAssembly (Wasm) differs from the other runtimes I discussed since it is not tied to any specific programming language. Its website describes it as *"a binary instruction format for a stack-based virtual machine."* Announced in 2015, it saw its first release in 2017. WebAssembly became a World Wide Web Consortium recommendation[2] in 2019. A first public working draft for version 2.0 of the specification was published in April 2022.

Initially, WebAssembly was designed as a memory-safe, sandboxed execution environment providing near-native code execution speed inside web browsers. However, the portability stemming from its use of a virtual machine made it attractive for other uses. There are now many general-purpose runtime implementations in the market that enable you to deploy WebAssembly code at the Edge or in the Cloud.

WebAssembly implementations rely on ahead-of-time (AOT) or just-in-time (JIT) compilation. However, some leverage an interpreter. At the time of writing, approximately 40 programming languages supported WebAssembly as a compilation target. The better-known options include C/C++, C#, F#, Go, Kotlin, Rust, and Swift. Independently of the original language used, WebAssembly code can be pretty-printed in a textual format defined by an attribute grammar.[3]

Since it is purely a runtime, the availability of libraries supporting your preferred IoT protocols will be determined by whichever programming language you select. Given developers' growing interest in deploying WebAssembly code outside the browser, there is a burgeoning ecosystem of orchestration platforms targeted at Edge and Cloud environments. However, at this point, none has the maturity of the leading platforms for virtual machine or container orchestration. WebAssembly is nevertheless a strong contender if you wish to build portable, high-performance applications. I strongly recommend you experiment with a few open source runtime implementations to find one that suits your needs.

How to Choose Your Runtime

All the options I discussed are good choices for building Edge and IoT Cloud applications with a robust ecosystem of open source libraries. You cannot go wrong with any of them. Ultimately, you should go with programming languages you

---

[2] https://www.w3.org/TR/2019/REC-wasm-core-1-20191205/

[3] You can see a good tutorial covering the topic on Mozilla's website: https://developer.mozilla.org/en-US/docs/WebAssembly/Understanding_the_text_format

are comfortable with and that make you productive. Naturally, the availability of developers with the required skill set is also a consideration. From a forward-looking standpoint, you should seriously look at Rust and WebAssembly if possible. The memory safety features of the former make it more secure than alternatives, while the latter provides application portability. Whatever option you choose, just ensure you rely on dependencies that are of reputable origin and at least receive timely security updates. Attention to IP and licensing issues will also pay off over the long term.

Web of Things

By nature, IoT and edge computing deployments are heterogeneous. I have made that point several times already. The devices and software platforms currently in the market involve a variety of protocols, data models, and security requirements. Given the complexity involved, developers and researchers have pursued the idea of connecting physical objects through web technologies since the early 2000s. These efforts intensified over time, leading to the creation of the Web of Things (WoT) working group at the World Wide Web Consortium (W3C) in 2016.

The Web of Things provides a set of standardized building blocks simplifying the development of IoT applications by leveraging the mature Web paradigm. Specifically, it establishes an interaction model composed of properties, actions, and events. Properties reflect the state of the "thing"; they can be sensor values, configuration settings, or the result of some computation. Properties can be observable, depending on the TD provider used. On the other hand, actions often involve interactions with the physical world. Typical examples would be powering a motor, opening a valve, or unlocking a door. Finally, events communicate the existence of use case–specific situations detected by the device, such as "low fuel," "valve opened," or "intruder detected." You can access the WoT architecture specification here: `www.w3.org/TR/wot-architecture11/`.

WoT is an open standard. It relies on many established web standards like HTTP, JSON, JSON-LD, JsonSchema, JsonPointer, and others. It is under active development by a diverse community. Like the Eclipse Foundation Specification Process, the W3C's specification process is transparent and open. Decisions are made publicly, meeting minutes are available to all, and the specifications are developed in public GitHub repositories.

The WoT model describes physical objects in a *WoT Thing Description* (TD). The TD is the central pillar of WoT. Since it provides access to a thing's metadata, you can consider it the equivalent of a website's homepage. The WoT TD describes which data and functions the thing provides, which protocol is used to access them, the structure and encoding of the data, and the access control mechanisms involved. The thing can also provide additional metadata in machine-readable or human-readable formats. You can retrieve a thing's TD from the device itself, a repository (TD Directory), or a third-party intermediary such as Eclipse Ditto. TDs are expressed in JSON-LD.

The next revision of the WoT Thing Description specification (version 1.1) will introduce the Thing Models (TM) concept. A TM defines a Thing's properties, actions, and events without providing instance-specific details such as the protocols or binding templates available. At the time of writing, the WoT Thing Description specification version 1.1 was available as a working draft.

WoT is protocol-agnostic. However, it provides a mechanism to define a mapping between the WoT's properties-action-event abstraction and specific protocol implementations such as CoAP, HTTP, and MQTT. The name of this mechanism is *WoT Binding Templates*. Binding templates define how the consumer of a Thing instance can activate the WoT interactions specified in a TD through the interface of a protocol.

In addition to TDs and binding templates, the WoT working group maintains a WoT Scripting API targeting JavaScript (ECMAScript). This API implements the properties-action-event abstraction and defines an interface scripting-based runtimes can leverage. It is, of course, possible to implement the same API in nonscripting languages such as Java, Python, or Rust. When I wrote this, the API was not a fully fledged WoT specification but was available as a Working Group Note. The working group also provides security and privacy guidelines for WoT implementations.

WoT provides useful abstractions enabling developers to implement connected objects. The Eclipse Thingweb project, which I will cover in detail later, is the reference implementation of the WoT Scripting API.

Eclipse EdiTDor

Written in JavaScript, Eclipse EdiTDor can create, render, edit, and validate WoT Thing Descriptions. It uses React for its UI framework and requires Node.js to build the web application. EdiTDor features wizards to create properties, actions, and events. It also ships with a JSON editor supporting validation through JSON schemas and auto-completion. EdiTDor provides two-way binding to the TM/TD format expressed as

JSON-LD, which provides a way to encode linked data using JSON. Figure 12-1 shows what EdiTDor looks like.

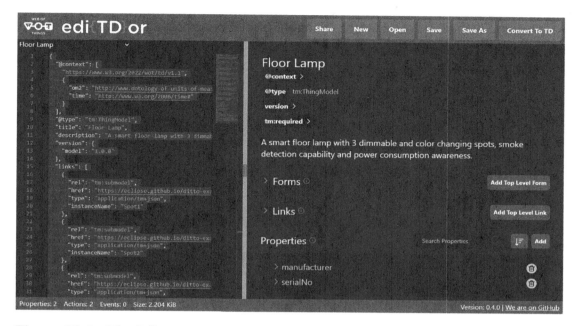

Figure 12-1. *The EdiTDor user interface*

Once you are happy with your Thing Models or Thing Definitions, one way to make them available to consumers is to deploy the files on an HTTP server. To that effect, you can save a copy of the TM or TD you created in EdiTDor.

Getting Started with EdiTDor

The EdiTDor team made an instance of the application available at the following location: `https://eclipse.github.io/editdor/`. Suppose you have Node.js version 10 or later installed on your workstation. In that case, you can also run a local instance effortlessly by cloning the Git repository of the project and executing the following command from the root folder:

```
yarn start
```

Once started, you can access it at the following URL: `http://localhost:3000/`.

Eclipse Thingweb

Eclipse Thingweb is a project providing software platforms implementing the WoT specification. At the time of writing, the only shipping platform was node-wot, the reference implementation of the WoT specification.

The official web resources for Thingweb are as follows:

- **Website:** `www.thingweb.io/`

- **Eclipse project page:** `https://projects.eclipse.org/projects/iot.thingweb`

- **Code repository:** `https://github.com/eclipse/thingweb.node-wot`

node-wot provides a WoT Thing Description parser and serializer, several binding template implementations (including HTTPS, CoAP, and MQTT, among others), and a runtime system conforming to the WoT Scripting API for applications. node-wot also includes a web user interface to visualize TDs and to enable interaction with Thing instances. node-wot is available under either the Eclipse Public License version 2.0 or the W3C Software Notice and Document License (2015-05-13). At the time of writing, the current version was v0.8.1, made available on May 19, 2022.

node-wot is written in JavaScript and uses the Node.js runtime. This means you can use it in three ways: as a library, a stand-alone application, or inside the browser. The latter restricts the binding templates available to HTTPS and WebSocket. node-wot can consume (client) and expose (server) Thing instances.

In addition to node-wot, the Thingweb team intends to work on a Thing Directory in the future. A Thing Directory is a directory service for WoT TDs that provides a service interface to register and look up for TDs. The Thingweb directory will conform to RFC 9176 (Constrained RESTful Environments (CoRE) Resource Directory).

Getting Started with node-wot

How to get started with node-wot depends on how you will leverage the platform. I will focus here on using it as a library or stand-alone application since those are the two most common use cases.

Since node-wot relies on Node.js, the team made the platform accessible as a dependency through the NPM package manager. Given node-wot's modularity, there are several packages available. The core platform is in node-wot-core, and there are separate packages for each binding template. You can list all the relevant packages by executing this command:

```
npm search @node-wot
```

Let's say all you want to use is the HTTP binding template. Then, you can install the relevant dependencies with this command:

```
npm i @node-wot/core @node-wot/binding-http –save
```

Once the libraries are available, you can build your application. The following code snippet shows how to instantiate a WoT Server.

```
const Servient = require('@node-wot/core').Servient;
const HttpServer = require('@node-wot/binding-http').HttpServer;

const servient = new Servient();
servient.addServer(new HttpServer());

servient.start().then((WoT) => {
    let thing = WoT.produce({
        title: "ThingTThing",
        description: " Beware of the Thing.",
    });

    // Add code to define properties, actions and
    // events here.

    thing.expose();
});
```

Once the server is up, the preceding code uses WoT.produce to create one Thing instance that will be available through the configured binding template. The title of the thing determines the URL. By default, the HTTP template will rely on port 8080.

Suppose you already started a server running the preceding code on your workstation. How would you consume the Thing instance you just created through HTTP? The following code snippet does just that:

```
const { Servient, Helpers } = require("@node-wot/core");
const { HttpClientFactory } = require('@node-wot/binding-http');
```

```
const servient = new Servient();
servient.addClientFactory(new HttpClientFactory(null));
const WoTHelpers = new Helpers(servient);

WoTHelpers.fetch("http://localhost:8080/ThingTThing").then(async (td) => {
    try {
        servient.start().then(async (WoT) => {

            let thing = await WoT.consume(td);
            // Use the thing's properties and actions here.
        });
    }
    catch (err) {
        console.error("Script error:", err);
    }
}).catch((err) => { console.error("Fetch error:", err); });
```

Naturally, what you can achieve with the Thing instance depends on what properties and actions it implements. The node-wot repository on GitHub contains an overview of the APIs involved in exposing and consuming Things. You can find it here: `https://github.com/eclipse/thingweb.node-wot/blob/master/API.md`.

Digital Twins

Up to now, I have discussed IoT applications in microservices and the runtime platforms supporting them. While this approach is fine, some higher-level abstractions have gained popularity in the last few years. The most prominent of these is digital twins.

At its core, a digital twin is a virtual model accurately reflecting a physical object's properties and state. The *concept* of digital twins was first expressed in 1991 by David Gelernter in a book titled *Mirror Worlds*. However, it was only applied to manufacturing in 2002 by Dr. Michael Grieves, a then faculty member at the University of Michigan. The *term* "digital twin" was eventually coined in 2010 by NASA's John Vickers. That said, NASA missions back to the 1960s already relied on the approach. The space agency built replicas of each spacecraft it launched, which were used for learning and simulation purposes. It is fascinating that space exploration was instrumental in the inception of edge computing (as described in Chapter 1) and digital twins.

> **Note**   Digital twins apply to many industries. Manufacturing, construction, healthcare, and automotive are some of the best examples. Organizations like the Industrial Digital Twin Association (IDTA) are working to define specifications for interoperability and integrate harmonized submodels. This led to the creation of the Digital Twin top-level project at the Eclipse Foundation.

Many products are proposing an approach for digital twins. In the remainder of this section, I will propose an approach leveraging existing standards and open source projects for their implementation.

Eclipse Ditto

Eclipse Ditto is a digital twin platform abstracting device interactions via their digital twins. It supports both device-to-cloud and cloud-to-device interactions. Using Ditto, you can build digital twins for new or existing devices; in other words, it supports both brownfield and greenfield deployments. As all device interactions are orchestrated through their digital twins, Ditto enforces authorization on all API calls. The platform also provides SDKs for interacting with digital twin instances through HTTP, WebSockets, and other widespread protocols such as MQTT and Apache Kafka. Ditto is explicitly designed as a Cloud back end. It can scale horizontally and is capable of providing digital twins for millions of connected devices at once.

Bosch contributed the Ditto project to the Eclipse Foundation in 2017. Ditto is licensed under the Eclipse Public License version 2.0. When I wrote this, the current version was v2.4.0, released on April 14, 2022.

The official web resources for Ditto are as follows:

- **Website:** https://eclipse.org/ditto

- **Eclipse project page:** https://projects.eclipse.org/projects/iot.ditto

- **Code repository:** https://github.com/eclipse/ditto

Ditto is a prime example of a well-thought-out separation of concerns. Its focus is only on providing a digital twin back end, leaving device connectivity and software updates to other specialized platforms. The functional scope of Ditto can be synthesized in this way:

- Providing APIs abstracting the details of the IoT protocols used by physical devices

- Routing requests between applications and devices or the device's last known state persisted in Ditto

- Enforcing authorization

- Persisting the last known values sent by the devices

- Notifying interested parties about changes in the persisted state of the digital twins

- Establishing and maintaining connections to message brokers, enabling asynchronous interaction with digital twins from third-party party applications

- Applying payload transformations when required to map the payloads of existing brownfield devices into a format Ditto can understand.

- Maintaining a search index of all persisted twin data

- Exposing APIs to process queries against sensor data of complete device fleets

There are a few core entities in Ditto. Here is a list of the main of those entities:

- **Thing:** Generic entity representing a device (physical or virtual), a transaction, master data, or anything that can be modeled. A Thing is the digital twin instance; its interaction capabilities can be defined, for example, through a WoT Thing Model. A thing possesses attributes and features. Attributes are static metadata, such as a serial number. Features reflect a thing's state; this can be sensor values or configuration settings, for example.

- **Message:** Arbitrary payloads sent to and from the entity represented by a Thing. Messages sent to a device are operations that should trigger an action. Messages received from a device are events propagated by the entity. Ditto does not perform message retention for offline entities. By default, each message will be delivered at most once, meaning there is no delivery guarantee. However,

Ditto provides mechanisms for "at least once" delivery through acknowledgments.

- **Policy:** Specifies fine-grained access control for Things, Features, Messages, and Policies themselves.

Internally, Ditto relies on a signal-based system to keep track of the digital twins (Things) and the entities they represent. Those signals are commands, command responses, error responses, events, and announcements. Ditto also manages acknowledgments that indicate that a signal was received or processed successfully. In addition, Ditto possesses the concept of connections, a separate entity with a different life cycle representing a channel used to exchange messages. Such connections can be established with device connectivity platforms like Eclipse Hono, MQTT brokers, or messaging services like RabbitMQ or Apache Kafka.

Ditto also defines a protocol of the same name, registered as an official IANA media type. This protocol relies on JSON payloads and can run over several network transport protocols, namely, AMQP (0.9 and 1.0), HTTP 1.1, MQTT (3.1.1 and 5.0), Kafka 2.x, and WebSocket. The payloads represent application data and are wrapped into an envelope used to route requests to their destination. To simplify the usage of the Ditto protocol, the project team maintains Java and JavaScript SDKs that you can use in your applications that need to create digital twins or interact with them. You can also use the Ditto REST and WebSocket APIs for the same purpose.

The Ditto platform is implemented in microservices communicating asynchronously through predefined signals. You can scale each of them horizontally depending on your scalability requirements. Here is the list of those services:

- **Policies:** Responsible for the persistence of Policies.

- **Things:** Responsible for the persistence of Things and their Features.

- **Things-Search:** Tracks changes to Things, Features, and Policies. This service also maintains a search index and executes queries against it.

- **Concierge:** Orchestrates the services that perform persistence and authorizes requests against them.

- **Gateway:** Exposes REST and WebSocket APIs.

- **Connectivity:** Sends and receives messages through the channel established to several external messaging systems.

Figure 12-2 shows how those microservices interact in the context of Ditto's architecture.

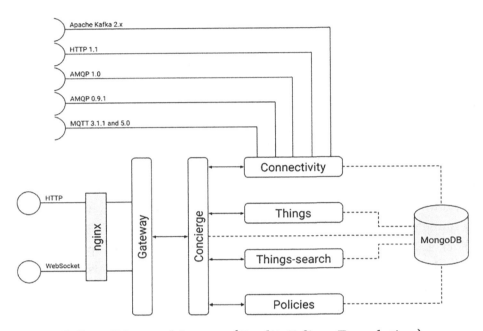

Figure 12-2. *Eclipse Ditto architecture (Credit: Eclipse Foundation)*

You probably noticed there are two third-party components featured in the architecture. One is Nginx, which is used as a reverse proxy, and the other is MongoDB, which is used for the persistence of Policies, Things, Connections, and the Things search index.

Getting Started with Ditto

Ditto requires a bit more resources than some of the other platforms covered in this book. The microservices are delivered through container images and run on various engines, including Docker, Kubernetes, and OpenShift. Podman could also be an alternative, given that it strives for Docker compatibility. At a minimum, the Ditto team recommends you allocate 2 CPU cores and 4 GB of memory to spin up a local instance on your workstation. You will probably need more resources if orchestrating the containers with a Kubernetes distribution. Alternatively, you can use the Ditto public sandbox in your experiments. It is reachable over HTTP and WebSocket at `ditto.eclipseprojects.io`.

The Ditto team provides installation instructions here: `https://github.com/eclipse/ditto/blob/master/deployment/README.md`. Alternatively, you could deploy the Cloud2Edge package from the Eclipse IoT packages project on a Kubernetes cluster. Cloud2Edge provides pre-configured and integrated instances of Eclipse Ditto and Eclipse Hono. I will cover Hono in the next chapter. For the time being, I will just explain how to deploy Ditto on its own using Docker on the Linux or macOS platforms.

The files related to Docker deployment are in the GitHub repository at `https://github.com/eclipse/ditto/tree/master/deployment/docker`. The easiest way to get them on your workstation is to clone the repository. Once you have done this, open a command-line prompt and go to the `deployment/docker` subfolder.

The first step is to configure security on the Nginx reverse proxy. The `nginx.conf` file provided by the Ditto team sets up Basic Authentication on both the HTTP and WebSocket endpoints. You need to assign a password to the `ditto` user and any account you want to add. To create such passwords, run the following command:

```
openssl passwd -quiet
 Password: <enter password>
 Verifying - Password: <enter password>
```

Once you confirm the password, the command will print a hash. You should paste that hash into the `nginx.httpasswd` file. The following is an example. You can create additional users by adding them and their hash to the file.

```
ditto:ZKrYJB4LjwQMQ
```

You are now ready to start your Ditto instance. Simply run this command to do so:

```
docker-compose up -d
```

To make sure everything is running properly, you can use this command:

```
docker-compose logs -f
```

Getting Started with the Client SDK

As I mentioned, the Ditto Client SDK is available for Java, JavaScript, Go, and Python. I will show you how to use the Java version here.

The Ditto team publishes its Java Client SDK on Maven Central. To add it to your project, all you need to do is to add the following dependency to your pom.xml:

```
<dependency>
    <groupId>org.eclipse.ditto</groupId>
    <artifactId>ditto-client</artifactId>
    <version>${ditto-client.version}</version>
</dependency>
```

New versions of the SDK are published in lockstep with Ditto releases. Keeping the two of them in sync is recommended, although the team strives to make the SDK forward compatible with the back end.

Before doing anything, you need to create and configure a Ditto client instance. The relevant classes are part of the org.eclipse.ditto.client.configuration package. The first step in the process is to create instances of AuthenticationProvider and MessagingProvider. For AuthenticationProvider, things look like this when Basic Authentication is used:

```
AuthenticationProvider authProvider =
    AuthenticationProviders.basic(
        BasicAuthenticationConfiguration.newBuilder()
        .username("ditto")
        .password("ditto")
        .build());
```

The MessagingProvider is similarly straightforward. I use the WebSocket URL for the public Ditto sandbox as the target endpoint in the following snippet:

```
MessagingProvider messProvider =
    MessagingProviders.webSocket(
        WebSocketMessagingConfiguration.newBuilder()
            .endpoint("wss://ditto.eclipseprojects.io")
            .build(), authProvider);

DittoClient client =
    DittoClients.newInstance(messProvider);
```

Once you have a client, you can manipulate Things and Policies. For example, to create a Thing instance and persist it on the back end, you would use the following code:

```
final Thing thing = Thing.newBuilder()
        .setId(ThingId.of(
                        "com.blueberrycoder:LaserShark-93110798"))
        .setFeature("Laser",
                FeatureProperties.newBuilder()
                    .set("Power", 9001)
                            .build())
        .build();
// It's over 9000!

client.twin().create(thing).get(1, SECONDS);
```

In the preceding example, I created a thing containing a single feature, itself made of a single property. The thing only persists on the Ditto back end when the create() method is called.

The Ditto team maintains an extensive set of examples for all SDKs, available in the ditto-examples repository. The repository can be found here: https://github.com/ eclipse/ditto-examples.

How It All Fits Together

EdiTDor, Thingweb node-wot, and Ditto can be used together in a WoT-centric workflow. I will now explain how. Basically, Ditto can create and manage digital twins without relying on WoT. However, when Ditto twin instances reference a WoT Thing Model, Ditto can generate a TD for them. A WoT consumer can access these TDs and the properties, actions, and events through the HTTP binding template. Naturally, the whole workflow relies on the availability of TMs. Since writing them by hand requires a deep understanding of the complete WoT TD and TM specification, members of the Eclipse community created the Eclipse EdiTDor graphical editor to streamline this task.

As a developer, you can thus use EdiTDor to create TMs. You can then make them available through an HTTP server. This will enable you to create Thing instances (twins) referencing the TMs through the Ditto SDKs or the platform's REST API. Ditto can generate TDs for the twin instances, which you can consume from node-wot or another WoT-compliant platform.

Here is an example. The Ditto team makes a TM describing a floor lamp available at the following URL:

```
https://eclipse.github.io/ditto-examples/wot/models/floor-lamp-
1.0.0.tm.jsonld
```

We can create a Ditto thing using the REST API with a simple `curl` request.

```
curl --location --request PUT -u ditto:ditto 'https://ditto.
eclipseprojects.io/api/2/things/com.blueberrycoder:floor-lamp-0813' \
--header 'Content-Type: application/json' \
--data-raw '{
    "definition": "https://eclipse.github.io/ditto-examples/wot/models/
    floor-lamp-1.0.0.tm.jsonld"
}'
```

If successful, the request will return a 201 response code, and you will see the Ditto definition for the Thing payload in JSON format as the response body.

At this point, it is possible to consume the Ditto Thing as a WoT TD (Thing Description) through the HTTP binding template, which is the only one Ditto supports for the time being. The following curl request will retrieve the TD definition of the Thing I just created:

```
curl --location --request GET -u ditto:ditto 'https://ditto.
eclipseprojects.io/api/2/things/com.blueberrycoder:floor-lamp-0813' \
--header 'Accept: application/td+json'
```

The request will return a 200 response code and the Thing's definition in WoT TD format as the response's body. This happens because I specified the appropriate MIME type (`application/td+json`) in the request.

At this point, you can consume the Thing instance from node-wot. The interest in doing this is that Ditto can host the Things that are digital twins in a wider system while node-wot brokers access real devices and other software components that are not twins.

A Few Examples of Applications

Digital twins and the runtimes I covered earlier provide ways to build enterprise IoT applications and platforms. You can then deploy the components of those applications in various locations across the Edge-to-Cloud continuum. In the remainder of this chapter, we will explore three applications or application frameworks that illustrate what you could build or run.

Eclipse 4diac

Eclipse 4diac provides an open source infrastructure for distributed industrial process measurement and control systems. It is based on the IEC 61499 standard, which defines a domain-specific modeling language based on function blocks for building distributed industrial control solutions. The project was announced in 2007 by Thomas Strasser and Alois Zoitl, and it was contributed to the Eclipse Foundation in 2015. The code is available under the Eclipse Public License version 2.0.

There are two major components in 4diac: Forte and IDE. Forte is a portable small-footprint implementation of a runtime compliant with IEC 61499. It is written in C++ and targets 16- and 32-bit constrained controllers. It can run on several operating systems, with Linux, VxWorks, and Windows among the better-known options. FreeRTOS support is in the works. From a protocol standpoint, Forte supports TCP and UDP over IP, as well as MQTT and OPC UA, among others. Using serial (RS-232) communications is also possible. IDE, on the other hand, is a set of plugins for the Eclipse Desktop IDE written in Java. It provides a design environment for developers targeting the Forte runtime.

Figure 12-3 shows 4diac IDE in action.

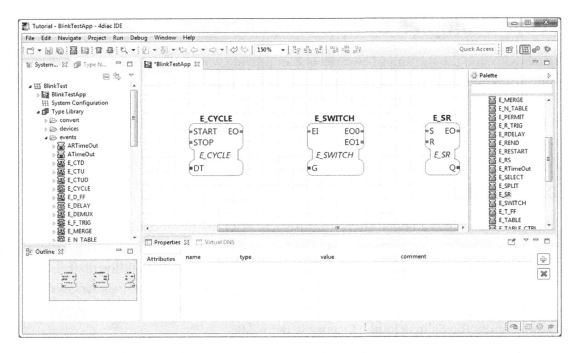

Figure 12-3. *4diac IDE (Credit: Eclipse Foundation)*

The official web resources for 4diac are as follows:

- **Website:** https://eclipse.org/4diac

- **Eclipse project page:** https://projects.eclipse.org/projects/
  iot.4diac

- **Code repositories:** https://git.eclipse.org/c/4diac/org.
  eclipse.4diac.forte.git/ (runtime)

  https://git.eclipse.org/c/4diac/org.eclipse.4diac.ide.
  git/ (IDE)

4diac IDE supports the development process by providing a visual editor to build
function block (FB) diagrams. IDE ships with several predefined function blocks, but you
can also design custom ones.

4diac Forte is usually deployed on programmable logic controllers (PLCs).
Those are ruggedized industrial computers that control manufacturing processes.
The project's documentation provides instructions to work with various hardware

platforms. Moreover, 4diac IDE sports a device simulator, enabling you to run and debug applications targeting Forte without needing specialized hardware.

Manufacturing and process control are usually highly distributed, with PLCs acting as edge nodes and local or remote Supervisory Control and Data Acquisition (SCADA) systems coordinating operations. Since it is written in C++, 4diac's runtime is the standard C library provided by the target operating system or compiler.

Eclipse Keyple

If you went to France and took an SNCF train or hopped on a subway or bus in Paris, you already used edge applications based on Eclipse Keyple without knowing it. Keyple is an open source framework facilitating the implementation of terminals supporting smart card–based processes and supporting secure ticketing transactions. Keyple was contributed to the Eclipse Foundation in 2018 by the Calypso Networks Association. It is published under the Eclipse Public License version 2.0.

The official web resources for Keyple are as follows:

- **Website:** https://keyple.org

- **Eclipse project page:** https://projects.eclipse.org/projects/iot.keyple

- **Index of code repositories:** https://github.com/eclipse/keyple

There are essentially two implementations of Keyple: one in Java and the other in C++. Both are roughly equivalent from a feature perspective. At the time of writing, the current version of the framework was v2.0, which is a completely modularized rewrite of the previous version. This is the version you should work with. Although the code base is now spread over several repositories, the project team put together tools and instructions to easily get the components your application requires.

The core features of Keyple are focused on smart card readers. Keyple provides plugins that support several reader interface standards, including PC/SC, NFC (on Android), and OMAPI. The readers can be local or remote, enabling the creation of highly distributed solutions at the Edge. Keyple also supports the Calypso processing API to manage Calypso commands and security features. This essentially means that your application can in real time determine a ticket's balance, confirm the ticket holder's privileges, and update the ticket data to deduce the journey cost and calculate a new balance, for example.

You will find Keyple-based applications in all sorts of settings at the Edge. They can handle all sorts of smart card–based use cases, not just ticketing. The Java version makes it easy to port your code to new hardware, while the C++ can support a broader assortment of readers and devices, including very constrained ones that could not support the footprint of the Java runtime.

Eclipse VOLTTRON

Eclipse VOLTTRON is an open source platform for distributed sensing and control. It provides services for gathering and storing data from devices and buildings, plus an environment for developing applications that interact with that data. VOLTTRON was created by a team at the Pacific Northwest National Laboratory (PNNL) and contributed to the Eclipse Foundation in 2019. It is licensed under the Apache License version 2.0.

The official web resources for VOLTTRON are as follows:

- **Website:** `https://volttron.org`

- **Eclipse project page:** `https://projects.eclipse.org/projects/iot.volttron`

- **Code repository:** `https://github.com/eclipse-volttron`

VOLTTRON is written in Python and targets the Linux operating system. The platform is made of several distinct components. Here is a list of the main components:

- **Agents:** Software components performing tasks on behalf of the users. Common use cases are data collection, device control, and platform management.

- **Message bus:** Backbone of the system. It allows agents to publish their results and messages and to subscribe to data sources. You can configure VOLTTRON to use ZeroMQ or RabbitMQ to provide the message bus.

- **Driver framework:** Interface between VOLTTRON agents and a device. The platform ships with a rich set of drivers, and you can develop your own.

- **Historian framework:** API agents can leverage to automatically collect data from a defined subset of topics on the message bus and persist them to a historian database. The project team currently maintains historians targeting SQL databases, MongoDB, and CrateDB, among others. You can naturally build your own.

- **User interface:** Web graphical management interface for managing distributed instances from a central one.

Figure 12-4 gives you a glimpse of the VOLTTRON user interface.

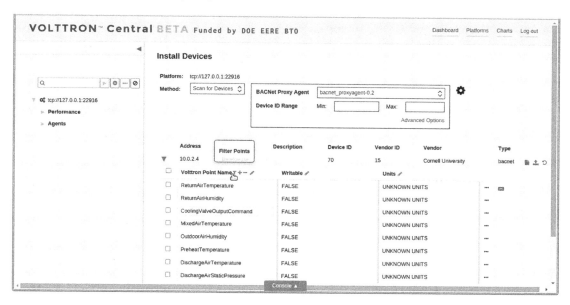

Figure 12-4. *VOLTTRON management user interface (Credit: Eclipse Foundation)*

VOLTTRON is a great example of a platform and framework supporting the development of distributed IoT applications. You can use it to build your own task-specific agents to cover many use cases. PNNL and its partners have successfully used it to build energy management solutions targeted at digital buildings.

CHAPTER 13

Integration and Data

Il ne faut pas uniquement intégrer. Il faut aussi désintégrer. C'est ça la vie. C'est ça la philosophie. C'est ça la science. C'est ça le progrès, la civilisation.

It's not enough to integrate, you must also disintegrate. That's the way life is. That's philosophy. That's science. That's progress, civilization.

—Eugène Ionesco, *La leçon*

IoT and edge computing infrastructure may be deployed outside the corporate data center or the Cloud, but it is a crucial part of the organization. Executives and process owners need data from constrained devices to make informed decisions. They need those decisions enacted by the field through actuators. Edge nodes play a crucial role in all of this since they help deliver the latency, resiliency, and data sovereignty that make possible the deployment of real-time, mission-critical applications. However, this would not be enough to make your IoT application an *enterprise* IoT application.

The gradual computerization of organizations started in the 1960s in most industrialized countries. Although both industrial and business processes were impacted, they evolved in distinct ways. Business processes became the realm of information technology (IT). IT focuses on data modeling and data objects expressed through standardized data formats; it adopted patterns such as publish/subscribe early and leveraged integrated networks. On the other hand, industrial processes belonged to operational technology (OT). OT focuses on applications that are directly coupled to processes, resulting in market-specific protocols and data formats, which are often proprietary. OT is often operated on isolated networks distinct from IT ones.

The distinct evolution of IT and OT led to a persistent gap between them. Enterprise IoT must bridge this gap to provide integrated device connectivity, management, and data integration. This chapter will explore Eclipse Foundation open source projects providing those capabilities.

© Frédéric Desbiens 2023
F. Desbiens, *Building Enterprise IoT Solutions with Eclipse IoT Technologies*,
https://doi.org/10.1007/978-1-4842-8882-5_13

Device Connectivity

Typical business applications do not understand IoT protocols. Even enterprise-focused runtimes such as Jakarta EE do not support protocols like MQTT or CoAP out of the box. How do you then connect constrained devices and edge nodes to business applications? Enter IoT device connectivity platforms.

The central feature of device connectivity platforms is to provide a simple and uniform API that business applications will utilize independently of the actual IoT protocol leveraged by the device. Such platforms will thus support most if not all the IoT protocols discussed in this book to broaden their appeal. Usually, they will enable devices to transmit sensor readings or other data. Business applications can also use them to invoke operations on devices.

In the Eclipse IoT ecosystem, two projects can fill the role of a device connectivity platform: Eclipse Hono and Eclipse Kapua. Since Kapua's feature set extends to device management, I will focus on Hono for now.

Eclipse Hono

Eclipse Hono is a pure device connectivity platform capable of supporting millions of devices simultaneously. The project was contributed to the Eclipse Foundation in 2016 by Bosch.IO, a Bosch subsidiary focused on IoT technologies. Hono is available under the Eclipse Public License version 2.0. The platform is written in Java and requires Java 17 or newer to compile and run.

The official web resources for Hono are as follows:

- **Website:** https://eclipse.org/hono

- **Eclipse project page:** https://projects.eclipse.org/project/iot.hono

- **Repository:** https://github.com/eclipse/hono

Hono's architecture is based on microservices. The platform defines a special type of microservice, called a *Protocol Adapter*, which maps a specific IoT protocol to Hono's APIs. Hono ships with protocol adapters for AMQP 1.0, CoAP, HTTP, and MQTT. You can also implement your own. In Hono parlance, there are two types of APIs in the platform: *northbound* APIs expose Hono to business applications, while the protocol adapters use *southbound* APIs. In Hono, all interactions happen through messages exchanged through a central messaging system. Most messages stream from the devices to the business applications; those messages are called *downstream* messages. Conversely, messages sent by business applications to devices are called *upstream* messages.

Hono defines three distinct messaging APIs:

- **Telemetry:** Messages sent through the telemetry API are usually sensor readings. They are sent frequently.

- **Event:** The event API is reserved for less frequent, more important messages. Hono can persist event messages and deliver them later if a business application is offline.

- **Command and Control:** This API is used by northbound applications to send messages conveying commands upstream to devices. Optionally, devices can return a response. A Command Router component transmits messages to the protocol adapters to which the target devices are connected.

Hono is inherently a multitenant platform. In other words, a single instance of the platform can manage devices belonging to multiple organizations simultaneously. Because of this, in addition to encrypted communications, Hono provides a cohesive security model where devices are authenticated and authorized through a device registry.

Hono's device registry exposes APIs to manage tenants, device registration, and credentials. By default, devices cannot connect to any protocol adapter. You must provision tenants and devices into the system before they can connect. In the case of devices, provisioning can be automated if authentication is performed through X.509 client certificates or if the devices connect to Hono through a gateway. There are two distinct implementations of the registry, each using a different database engine for persistence. One uses the MongoDB NoSQL database. The other can use any database for which a JDBC driver is available, with H2 and PostgreSQL being the options officially supported by the project team. Please note that for the time being, the JDBC-based registry does not support searching for devices.

Since it is designed to support highly scalable Cloud services, Hono can enforce limits for resource consumption per tenant for each protocol adapter you deploy. You can define limits for the number of connections, the duration of connections, and the volume of data published over a defined time interval. It is also possible to define a maximum number of devices registered per tenant and a maximum number of credentials per device.

Figure 13-1 shows how Hono's components fit in its architecture. For illustration purposes, the instance shown has the HTTP and MQTT protocol adapters running.

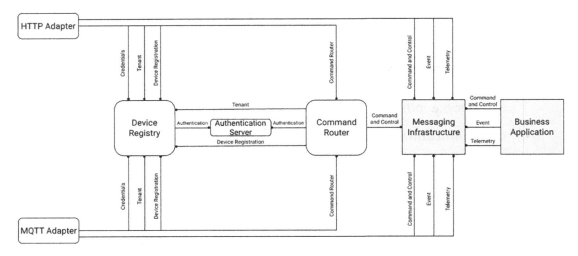

Figure 13-1. *Eclipse Hono architecture (Credit: Eclipse Foundation)*

One important precision about Hono is that it can work with two different types of messaging infrastructure. The first option is AMQP 1.0, using Apache Qpid Dispatch Router and Apache ActiveMQ Artemis. The example deployment currently uses a single Qpid Dispatch Router instance connected to a single Apache ActiveMQ Artemis broker. The second one is Apache Kafka. The example deployment currently leveraged Bitnami's Kafka Helm chart for installing a single-node Apache Kafka broker instance. Both setups are suitable for development purposes only. Scaling one of those options for production workloads is not something Hono can do for you.

Getting Started with Hono

To work with Hono, you will need at a minimum access to an instance of the platform. The project team maintains a publicly available sandbox that you can use. Naturally, you can also install a local instance. The platform is made of several containerized microservices. The Hono team provides a Helm chart to help you install them on a Kubernetes cluster. This chart has been developed in the context of the Eclipse IoT Packages project and is available at `https://github.com/eclipse/packages/blob/master/charts/hono/README.md`. If you do not have access to a Kubernetes instance, you can set up one yourself. The Hono team provides instructions to deploy a single-node cluster inside a local VM using Minikube here: `www.eclipse.org/hono/docs/deployment/create-kubernetes-cluster/`.

The Hono CLI utility is another useful tool you should have on your workstation. It can interact with messaging infrastructure implementing Hono's northbound Telemetry, Event, and Command and Control APIs. Additionally, it can interact with the AMQP protocol adapter. You can download it from this page on the Hono website: `www.eclipse.org/hono/downloads`. As an alternative, you can use `wget` if it is available on your operating system. Make sure you adjust the following URL to the latest version of the utility. It was 2.0.1 when I wrote this.

```
wget "https://www.eclipse.org/downloads/download.php?file=/hono/hono-cli-2.0.1-exec.jar&r=1" -O hono-cli-2.0.1-exec.jar
```

Since it is offered as an executable JAR, the CLI needs a Java runtime to run. Once the download is complete, you can test the JAR by running this command:

```
java -jar hono-cli-2.0.1-exec.jar -V
```

You should get output resembling this:

```
hono-cli 2.0.1
running on Linux 5.10.102.1-microsoft-standard-WSL2 amd64
```

In the rest of this section, I will demonstrate a few basic interactions against the Hono sandbox. You will find a fully fleshed-out tutorial here: `www.eclipse.org/hono/docs/getting-started/`

The Hono sandbox uses custom TCP ports; you could be prevented from accessing it if working from a corporate network. To validate that your workstation can reach the sandbox properly, execute the following request on the command line:

```
curl -sIX GET http://hono.eclipseprojects.io:28080/v1/tenants/DEFAULT_TENANT
```

The command should return output like this if your network administrators did not block the ports:

```
HTTP/1.1 200 OK
etag: aee7809e-ca66-4e5d-b722-f2f2c5661f87
```

```
content-type: application/json; charset=utf-8
content-length: 445
```

You should set up your own local instance if you get anything else.

The next step is to create a file used to set the required shell environment variables. Usually, it is named hono.env. When working with the sandbox, you can use the following command to create it:

```
cat <<EOS > hono.env
export REGISTRY_IP=hono.eclipseprojects.io
export HTTP_ADAPTER_IP=hono.eclipseprojects.io
export MQTT_ADAPTER_IP=hono.eclipseprojects.io
export KAFKA_IP=hono.eclipseprojects.io
export APP_OPTIONS="--sandbox"
EOS
```

Since Hono is a multitenant platform, you need to create your own tenant on the sandbox before being able to do anything else. Fortunately, the Hono Device Registry exposes a REST API for that purpose. You can invoke it with the following command:

```
source hono.env
curl -i -X POST -H "content-type: application/json" --data-binary '{
  "ext": {
    "messaging-type": "kafka"
  }
}' http://${REGISTRY_IP}:28080/v1/tenants
```

If all goes well, you will get a 201 response code and output resembling this:

```
HTTP/1.1 201 Created
etag: c11c7413-e019-4ef1-b984-5d8244aa1b2f
location: /v1/tenants/6e3eec20-3795-4f35-812a-4b45dabaffdf
content-type: application/json; charset=utf-8
content-length: 45
```

```
{"id":"6e3eec20-3795-4f35-812a-4b45dabaffdf"}
```

The JSON payload in the response body contains your unique randomly generated tenant ID. You need to add it to your shell environment file. For example:

```
echo "export MY_TENANT=6e3eec20-3795-4f35-812a-4b45dabaffdf" >> hono.env
```

Substitute the value you see here with the one returned by the preceding curl request.

The next step is to create a device and assign it a password. Don't worry; you do not need a physical device to complete the steps. The Hono CLI utility can consume the messages sent by devices and play the role of a virtual device as well.

To add a device to your tenant, execute this `curl` request:

```
source hono.env
curl -i -X POST http://${REGISTRY_IP}:28080/v1/devices/${MY_TENANT}
```

Once again, the response's body will contain a JSON structure providing the ID of the newly created device. In my case:

```
{"id":"53a3b746-d138-4f3c-8087-b580e241c9d2"}
```

You need to add this ID to your hono.env file as shown here:

```
echo "export MY_DEVICE=53a3b746-d138-4f3c-8087-b580e241c9d2" >> hono.env
```

The Hono sandbox requires authentication. You must define and assign a password to your device. Of course, Hono also supports client certificates for production deployments; I just use password authentication for the sake of simplicity here.

Pick a password and add it to your hono.env file like this:

```
echo "export MY_PWD=requin!" >> hono.env
```

Then, assign the password to the device through this REST API call:

```
source hono.env
curl -i -X PUT -H "content-type: application/json" --data-binary '[{
  "type": "hashed-password",
  "auth-id": "'${MY_DEVICE}'",
  "secrets": [{
      "pwd-plain": "'${MY_PWD}'"
  }]
}]' http://${REGISTRY_IP}:28080/v1/credentials/${MY_TENANT}/${MY_DEVICE}
```

The response code should be 204 (Updated), and the response body empty.

You are now ready to start your virtual device. Open a command-line prompt and navigate to the folder where you downloaded the Hono CLI. Then, execute the following command:

```
java -jar hono-cli-*-exec.jar app ${APP_OPTIONS} consume --tenant
${MY_TENANT}
```

If you did everything correctly, you should see output like this:

```
Connecting to Kafka based messaging infrastructure [hono.eclipseprojects.
io:9092,hono.eclipseprojects.io:9094]
Consuming messages for tenant [6e3eec20-3795-4f35-812a-4b45dabaffdf],
ctrl-c to exit.
```

Now, let's make an API call simulating a device sending telemetry. In this case, the payload is a simple JSON structure containing a single temperature value. Open a second command prompt and execute this command:

```
source hono.env
curl -i -u ${MY_DEVICE}@${MY_TENANT}:${MY_PWD} -H 'Content-Type:
application/json' --data-binary '{"temp": 5}' http://${HTTP_ADAPTER_
IP}:8080/telemetry
```

You should see messages like this in the shell where the Hono CLI is running:

```
t 53a3b746-d138-4f3c-8087-b580e241c9d2 application/json {"temp": 5}
{orig_adapter=hono-http, qos=0, device_id=53a3b746-d138-4f3c-8087-
b580e241c9d2, creation-time=1660328357552, traceparent=00-fbc396794954547a
68db4c1e0a6b8831-cfd5b098412f5788-01, content-type=application/json,
orig_address=/telemetry}
```

The Hono sandbox offers an MQTT protocol adapter as well. I will now show you how to simulate an MQTT device receiving a command.

To receive the command on your workstation, you must subscribe to the relevant MQTT topic. The easiest way to do this is to use the mosquitto_sub CLI utility shipped by the Mosquitto team. Please refer to the chapter on MQTT if you need to install it.

Run the following commands on a brand-new command prompt to start the MQTT subscription:

```
source hono.env
mosquitto_sub -v -h ${MQTT_ADAPTER_IP} -u ${MY_DEVICE}@${MY_TENANT} -P
${MY_PWD} -t command///req/#
```

You can stop the previously started Hono CLI with the Ctrl+C key combination. On the same prompt, issue this command:

```
source hono.env
java -jar hono-cli-*-exec.jar app ${APP_OPTIONS} command
```

This will start the Hono CLI in command mode, where you can issue device commands through a command interpreter. You should see this prompt:

```
hono-cli/app/command>
```

Execute this command, which simulates setting the volume to 11 on a remote sound system:

```
ow --tenant ${MY_TENANT} --device ${MY_DEVICE} -n setVolume --payload
'{"level": 11}'
```

If Hono successfully routes the command, you will see this in the shell where you are running mosquitto_sub:

```
command///req//setVolume {"level": 11}
```

Nigel Tufnel would be proud!

Device Management

When deploying solutions in the field, connecting IoT devices and Edge nodes to enterprise applications is only the first step. You will need to monitor them, update their software, and, over time, adjust their settings and configuration. There are, of course, several ways to achieve this. Some IoT protocols, such as LwM2M, possess built-in support for firmware updates. However, device management building blocks with a much broader scope are available in the open source ecosystem. In this section, I will focus on two of them: Eclipse hawkBit and Eclipse Kapua.

Eclipse hawkBit

Eclipse hawkBit is a domain-independent framework for rolling out software updates to IoT devices, edge nodes, and gateways connected to IP-based networks. The project was contributed to the Eclipse Foundation by Bosch.IO in 2015. hawkBit is written in Java and leverages the Spring framework and Spring Boot. Please note that counter to many other Eclipse Foundation projects, the code is published under the Eclipse Public License version 1.0 and not version 2.0.

The official web resources for hawkBit are as follows:

- **Website:** `https://eclipse.org/hawkbit`

- **Eclipse project page:** `https://projects.eclipse.org/project/iot.hawkbit`

- **Repository:** `https://github.com/eclipse/hawkbit`

The core concept of hawkBit is the *software module*. A module comprises one or several *artifacts*, that is, files that will be deployed on the devices. hawkBit lets you group software modules in *distribution sets* as needed. You can specify the targets explicitly or select them through a target filter when performing a deployment.

Like Eclipse Hono and Eclipse Ditto, hawkBit is made of several microservices. The feature set of the platform is focused yet comprehensive. Here is a list of hawkBit's main features:

- **Device and Software Repository:** hawkBit maintains separate lists for the devices that are provisioning targets and for the software distributions that can be deployed. It is possible to pre-commission devices; this is useful to support zero-touch deployment scenarios. hawkBit maintains a full history of software updates for all devices in the repository.

- **Software Updates Management:** hawkBit can deploy software distributions to devices. This action can be initiated through the web user interface or a REST API. Interestingly, hawkBit supports the download resumption mechanisms initially defined in RFC 7233. Devices can only access the distributions they are authorized to. You can deploy delta updates to devices, meaning that the artifact contains only the differences between two specific software versions.

The integrity of artifacts can also be attested to through digital signatures.

- **Rollout and Campaign Management:** hawkBit provides ways to handle large-scale deployments, also called *rollouts*. If you have hundreds or thousands of devices to update at once, you need to ensure your hawkBit instance and networking infrastructure can handle the load. Moreover, deploying software updates to subsets of your device fleet maintains a certain level of service and ensures you will not lose all your capacity at once in the case of a failed deployment. To perform a rollout, you need to split your device fleet into *deployment groups*; alternatively, hawkBit can generate those according to a predefined group size. During a rollout, the platform will process one group at a time in a cascade; the rollout will abort if the error rate for the group is over the threshold you provided.

hawkBit provides a fully featured web interface enabling you to use most of the preceding features. Figure 13-2 gives you a glimpse of how it looks.

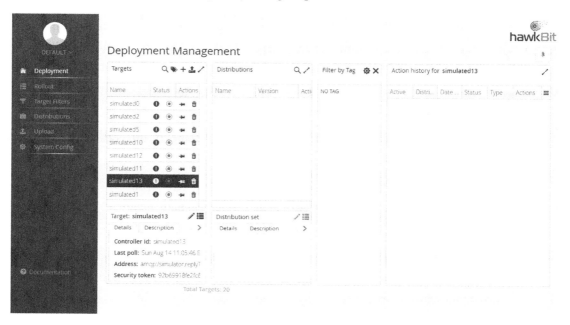

Figure 13-2. *The Eclipse hawkBit user interface*

From an architecture standpoint, hawkBit exposes three distinct APIs to consumers. Here are the main characteristics of each of those APIs:

- **Management API:** A REST API enabling you to create, read, update, or delete (CRUD) provisioning targets (devices) and repository content (software).

- **Direct Device Integration (DDI) API:** A REST API enabling device interactions. It uses JSON payloads and provides a channel for devices to send feedback to the platform. Naturally, this is useful for monitoring the progress of deployments, for example.

- **Device Management Federation (DMF) API:** An AMQP API enabling hawkBit to manage software updates for devices under the control of another device management service or application.

Documentation for each API is available on the hawkBit website at this location: `www.eclipse.org/hawkbit/apis/`.

There are several client implementations for hawkBit's Direct Device Integration (DDI) API. The best known comes from the Eclipse Hara subproject, which focuses on a Kotlin library called hara-ddiclient. Hara can be harnessed from any language running supported by the Java virtual machine (JVM). The subproject was contributed to the Eclipse Foundation in 2019 by Kynetics, a startup from California. Hara is available under the Eclipse Public License version 2.0. There are also a few non-Eclipse DDI client libraries. The main ones of those libraries are as follows:

- **SWupdate:** A Linux-based agent focusing on updates for embedded systems. See `https://github.com/sbabic/swupdate`.

- **rauc-hawkbit-updater:** A hawkBit client for the RAUC update framework written in C. See `https://github.com/rauc/rauc-hawkbit-updater`.

- **rauc-hawkbit:** A hawkBit client demo application and library for the RAUC update framework written in Python. See `https://github.com/rauc/rauc-hawkbit`.

- **hawkbit-rs:** A project providing Rust crates to help implement and test hawkBit clients. See `https://github.com/collabora/hawkbit-rs`.

From a code point of view, hawkBit depends on several third-party software components and platforms. Figure 13-3 shows where they fit in the overall architecture.

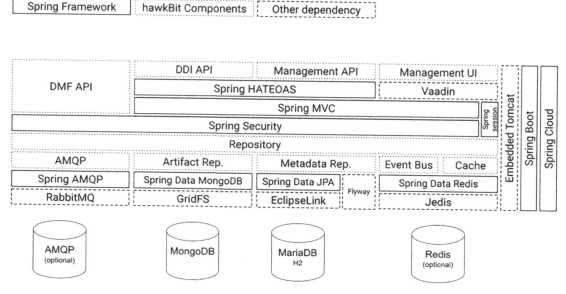

Figure 13-3. *Eclipse hawkBit architecture (Credit: Eclipse Foundation)*

The project team supports several choices for the SQL database engine. At the time, MySQL, MariaDB, and Microsoft SQL Server were the choices deemed suitable for production deployments. You could also use H2, IBM DB2, or PostgreSQL in development and test environments.

Getting Started with hawkBit

The easiest way to kick the tires of hawkBit is to use the public sandbox maintained by the project team. You can access it at `https://hawkbit.eclipseprojects.io`. You can also spin up a local environment by using the container images put together by the project team. If you wish to just start the core engine itself under the Docker engine, you can use this command:

```
docker run -p 8080:8080 hawkbit/hawkbit-update-server:latest
```

For an environment closer to a production one, you can start MySQL and RabbitMQ containers alongside hawkBit itself through Docker compose. Use these commands:

```
git clone https://github.com/eclipse/hawkbit.git
```

```
cd hawkbit/hawkbit-runtime/docker
docker-compose up -d
```

Finally, you can leverage Docker swarm to bring up the hawkBit Device Simulator and the Update Server, MySQL, and RabbitMQ. Simply run the following commands:

```
git clone https://github.com/eclipse/hawkbit.git
cd hawkbit/hawkbit-runtime/docker
docker swarm init
docker stack deploy -c docker-compose-stack.yml hawkbit
```

The simulator supports the DDI and DMF APIs and can simulate the existence of up to 4000 devices.

Eclipse Kapua

Eclipse Kapua is a modular IoT cloud platform to manage and integrate devices and their data. It consolidates device connectivity and management on a single platform while providing out-of-the-box integration with gateways based on Eclipse Kura. Kapua was contributed to the Eclipse Foundation by Eurotech in 2016, and the code is available under the Eclipse Public License version 2.0. At the time of writing, the current version was v1.6.1, published on July 1, 2022. What a gift for Canada Day!

The official web resources for Kapua are as follows:

- **Website:** www.eclipse.org/kapua/

- **Eclipse project page:** https://projects.eclipse.org/projects/iot.kapua

- **Repository:** https://github.com/eclipse/kapua

Here is a list of Kapua's main features:

- **Device Connectivity:** Field devices can connect to Kapua using MQTT over TCP/IP or WebSockets. The device connectivity service performs authentication and authorization while maintaining a device registry.

- **Device Management:** Kapua can perform various remote operations on connected devices, independently of the software stack they run. This includes managing the devices' configuration, services, and applications. For example, you can start or stop services and install, update, or remove applications. You can also execute operating system commands remotely and provision an initial configuration for a device.

- **Data Management:** Telemetry sent by devices is persisted to an Elasticsearch database and fully indexed.

- **Security:** Kapua's security model spans tenants, accounts, and users. In Kapua, accounts support a hierarchical access control structure. The platform also implements Role-Based Access Control (RBAC) for user identities, which means permissions are granted according to the "least privilege" principle.

- **Application Integration:** Kapua exposes a comprehensive REST API that third-party applications can use to interact with the platform and the devices connected to it.

Kapua also offers a web graphical user interface that administrators can use to perform device and data management operations. Figure 13-4 shows you what it looks like.

Figure 13-4. *Eclipse Kapua web user interface*

As shown in the figure, Kura allows you to schedule batch jobs and define custom tags.

Getting Started with Kapua

Kapua is offered as a set of container images you can deploy on Docker or OpenShift, including the lightweight MiniShift flavor. You can deploy them on any x86_64 system. Make sure to allocate at least 6 GB of memory and 2 CPU cores to ensure Kapua performs adequately.

To get started on Docker, execute the following commands on Linux or macOS:

```
git clone https://github.com/eclipse/kapua.git kapua
cd kapua/deployment/docker/unix
./docker-deploy.sh
```

You can also perform the deployment using a PowerShell prompt on Windows. Use these commands to do so:

```
git clone https://github.com/eclipse/kapua.git kapua
cd kapua\deployment\docker\win
./docker-deploy.ps1
```

I confirmed the aforementioned also works with Podman on Linux if you install the additional podman-docker package and manually get a copy of the docker-compose CLI. For example:

```
curl -SL "https://github.com/docker/compose/releases/download/v2.9.0/
docker-compose-linux-x86_64" -o docker-compose
```

Make the file executable and place it in an appropriate folder on your PATH; things will work properly.

Once the containers run, you can access the Kapua console at http://localhost:8081/ using kapua-sys as the user and kapua-password as the password to log in. You can find additional information about the containers, the ports they expose, and credentials for the various components of the platform at this location: https://github.com/eclipse/kapua/tree/develop/deployment/docker.

Data

Data is at the core of any IoT deployment. Ultimately, you are deploying sensors to gather data; your devices in the field will interact with the physical world because they will receive commands triggered by data. Given this, most IoT devices produce a great quantity of data and will produce even more in the future. For example, an IDC forecast from 2019 stated, "*The Growth in Connected IoT Devices Is Expected to Generate 79.4ZB of Data in 2025.*"[1] Whether this prediction will come to pass or not is irrelevant. IoT is about Data. One could argue that the realm of data is where information technology and operational technology have the greatest chance to meet.

From a technical standpoint, there is nothing different about managing IoT data compared to enterprise data. You can use SQL, NoSQL, or other types of databases, and the way to scale them is roughly like what you typically see in enterprise IT. The real choice is not about the query language but whether you must process data at rest or in motion. Traditional databases handle data at rest. To process data in motion, you rather need to leverage stream processing platforms. In the open source world, this usually means paying a visit to the Apache Foundation. The Apache ecosystem contains several relevant projects such as Apache Spark Streaming, Apache Flink, Apache Kafka, and Apache Storm. Organizations mainly use streaming platforms to support real-time decision-making and fulfill low-latency requirements.

Explaining which of the Apache platforms I mention to use would probably require a book of its own. That said, you probably noticed that Eclipse Hono and Eclipse Ditto integrate out of the box with Kafka. This likely represents a good starting point. Other Eclipse projects provide different integration points with Enterprise applications. For example, the Eclipse Amlen MQTT broker offers a bridge to messaging systems compatible with the Java Message Service (JMS) specification.[2]

In many cases, IoT deployments rely on multiple databases at a time to address different requirements. For example, raw sensor data is often stored in a historical database for safekeeping. Many organizations will leverage such databases as a data source for machine learning and analytics.

---

[1] See www.businesswire.com/news/home/20190618005012/en/The-Growth-in-Connected-IoT-Devices-is-Expected-to-Generate-79.4ZB-of-Data-in-2025-According-to-a-New-IDC-Forecast

[2] https://www.eclipse.org/amlen/docs/Developing/devjms_working.html

Note   Historian databases are a type of time-series database developed specifically to store operational process data. Compared to other types of databases, they provide enhanced data capture, validation, and aggregation capacities. They typically sport built-in data compression. Most industrial automation specialists use the term "tag" to designate a stream of process data. This refers to the physical tags placed on instrumentation to capture data manually before computers were used to automate industrial processes. As the name suggests, the records in historian databases are not updated after the initial insertion.

The high level of granularity of the data gathered by most IoT devices is often unnecessary outside the factory. Consequently, an important part of the design of any IoT solution is to figure out which values are meaningful from a business perspective and what is the appropriate refresh rate for them.

The Eclipse Foundation ecosystem contains a few projects that address data management and analytics. The best known is Eclipse Streamsheets, which I will now cover in detail.

Eclipse Streamsheets

Eclipse Streamsheets is a platform providing server-based spreadsheets that consume, process, and produce live data streams. The project was contributed to the Eclipse Foundation in 2020 by Cedalo, a German startup that also supports the development of the Eclipse Mosquitto MQTT broker. Streamsheets is written in JavaScript and runs on the Node.js runtime. The code is published under the Eclipse Public License version 2.0.

The official web resources for Streamsheets are as follows:

- **Website:** `https://docs.cedalo.com/Streamsheets/2.5/installation/`

- **Eclipse project page:** `https://projects.eclipse.org/projects/iot.Streamsheets`

- **Repository:** `https://github.com/eclipse/Streamsheets`

Streamsheets can consume data streams independently of their origin. By default, the open source edition can connect using the following protocols: MQTT, REST, and SMTP/IPOP3 (email). Streamsheets can also connect to MongoDB databases and

Apache Kafka instances. Additional options are available in Cedalo's commercially supported edition.

In Streamsheets, data flows are processed, analyzed, and visualized entirely through no-code spreadsheets called, you guessed it, Streamsheets. Naturally, you can rely on an array of common spreadsheet formulas to model and format the data. The spreadsheets are recalculated whenever a new message arrives; you can also refresh them on a schedule. Consequently, the conditions and formulas found in the spreadsheets are reevaluated frequently, which means that the visualizations they contain are updated in real time. If desired, it is possible to pause the spreadsheet to keep it frozen at a specific moment.

While easy to use, Streamsheets offers comprehensive security. It can connect to streams over TLS when the underlying protocol supports it and defines user roles that protect the integrity of the spreadsheets. Users with the developer role can create new streams and spreadsheets; users with the viewer role can only view what others have built. Of course, there is an admin role with access to everything.

Streamsheets' user interface mimics popular web spreadsheet services and will be easy to pick up by users of desktop spreadsheet software as well. Figure 13-5 captures Streamsheets' user interface displaying a live spreadsheet.

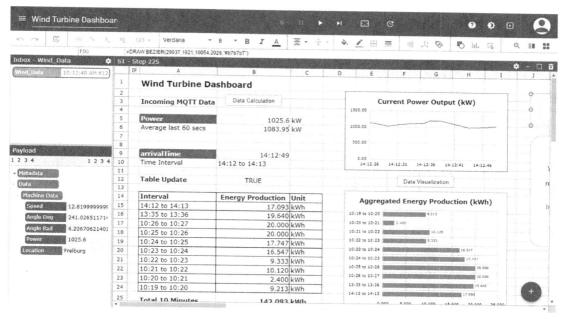

Figure 13-5. *Eclipse Streamsheets user interface*

Getting Started with Streamsheets

Streamsheets is offered as a set of containers that can run everywhere the Docker engine is supported. The project team recommends a quad-core processor and 4 GB of RAM for acceptable performance. Container images are also available for the Raspberry Pi. The minimum requirement is a model 3B equipped with 1 GB of RAM and 4 GB of storage; the recommended configuration is a model 4B with 2 GB of RAM and at least 10 GB of storage. Your Pi needs to run a 32-bit operating system.

Getting the containers up and running is easy. First, you need to execute the Cedalo installer. On Linux, execute this command:

```
docker run -it -v ~/cedalo_platform:/cedalo cedalo/installer:2-linux
```

The command looks like this for macOS:

```
docker run -it -v ~/cedalo_platform:/cedalo cedalo/installer:2-macos
```

Finally, this is the command on Windows:

```
docker run -it -v C:\cedalo_platform:/cedalo cedalo/installer:2-win
```

In all three cases, you will get a prompt like this one:

```
? Select what to install › - Space to select. Return to submit
◉    Management Center for Eclipse Mosquitto
◉    Eclipse Streamsheets
◉    Eclipse Mosquitto 2.0
◯     Eclipse Mosquitto 1.6
```

At a minimum, you should install Mosquitto 2.0 and Streamsheets. The Management Center will be useful if you wish for a powerful graphical interface for the Mosquitto broker but can be removed for resource-constrained environments. Move the cursor to whichever component you do not want to install and press space to remove it. Press return (enter) when you are ready to start the installation.

After a few minutes, the container images should be on your machine. Behind the scenes, the Cedalo installer relies on Docker Compose. Once the command prompt returns, you can move to the `cedalo_platform` folder and execute the startup script. On Linux and macOS, you need to issue the following commands:

```
cd ~/cedalo_platform
sh start.sh
```

On Windows, the commands are rather

```
cd C:\cedalo_platform
start.bat
```

As soon as the containers run, you can access Streamsheets at the following URL: http://localhost:8081. The default credentials are admin for the user and 1234 for the password. You should change the latter as soon as possible.

CHAPTER 14

Conclusion

It is September 2022. A few weeks ago, I wrote the last words of the last chapter of this book. So many things have happened in technology since I started to write in November 2021. The cadence of innovation is relentless. There was never a better time to work in IT. Technology is now deeply embedded in our lives, and this trend will only grow over time.

From a historical perspective, technological progress greatly improved our lives; machines made us more productive while reducing the time and effort required by various tasks. However, this increase in productivity came at a price: pollution. The mad race to extract and transform enough resources to sustain the increasing levels of industrial production needed to meet our ever-growing desires took a serious toll on our planet. Whatever lies ahead in the future, the days of perpetual economic growth are getting closer to an end – if they are not already over.

In this context, natural resources and our own time will be increasingly scarce and valuable. IoT has the potential to alleviate this scarcity. Sensors and actuators embedded in connected everyday objects have the potential to optimize our use of resources and free up our time. Smart agriculture means we can use just the right amount of water and fertilizer to optimize the yield of our crops. Smart buildings can use machine learning to anticipate occupation patterns in offices or classrooms and modulate heating and cooling accordingly, resulting in significant energy savings. Industrial automation can make factories more efficient and prolong the life of key equipment through preventive maintenance and real-time monitoring. All this potential, however, will be for naught if we don't address three acute issues: security, privacy, and short-sightedness.

Security has been, for a long time, the Achilles heel of IoT. There are many reasons for that. One is that it is simply hard to get right, given the complexity of IoT solutions. Another is that some device and software makers are simply obsessed with cost; consequently, they will ship products without proper support or the infrastructure and processes required to update them. Another possible reason is a lack of awareness or

F. Desbiens, *Building Enterprise IoT Solutions with Eclipse IoT Technologies*,
https://doi.org/10.1007/978-1-4842-8882-5_14

even plain inertia. Personally, the most problematic reason is the lack of investment in open source projects that became critical infrastructure. Think about OpenSSL and Log4j. Both projects were underfunded and understaffed; it was only when widespread vulnerabilities hit them that the industry deigned to dedicate a sliver of its profits to support them. So my advice to you is this: invest in your product's quality and future. Build value-driven rather than cost-driven solutions, and support the open source projects playing a central role in your infrastructure with money or development hours.

Privacy is the second issue. Consumers and organizations expect their data to be used for their own benefit, not yours. Yet the IT industry's obsession with the monetization of user data seems to be limitless. In other words, many technology suppliers are addicted to the easy money that can be made by showing you an endless stream of ads. But are these ads effective? From my point of view, those ads are an insidious form of pollution, preventing us from focusing. They are the background noise of our times. Don't believe me? Fire your favorite browser, and look for photographs of hockey games from the 1950s and 1960s. Not a cell phone nor an ad in sight. But gradually, ads made their way everywhere. They are on the boards; they are painted on the ice. They are blasted on the radio and TV at every break in the play; sometimes, referees need to wait until the end of commercials before restarting the game. More recently, they took over the names of buildings and even the uniforms themselves. But look at Figure 14-1. Look at all the ads on this player. Can you focus on anything? Remember any sponsor name after closing your eyes for a few seconds?

Figure 14-1. *A billboard on ice[1]*

So my advice to you is this: defend user privacy ferociously. And if you are thinking about a new product or business, think about a better business model than selling ads. The world doesn't need yet another AI-enabled money funnel pretending it will change our lives. Remember. Contrarily to open source contributors, ads do not generate value; they just send the money into a different pocket.

Finally comes the issue of short-sightedness. A few weeks before I wrote this conclusion, Google announced they would discontinue their IoT Core service in 2023. The platform's customers now need to scramble to find an alternative. This is one in a long string of such decisions stranding end users and developers by the search giant.

---

[1] By Pierre-Yves Beaudouin/Wikimedia Commons, CC BY-SA 3.0, `https://commons.wikimedia.org/w/index.php?curid=24733570`

In the past, people have even been left with broken, irreparable smart homes. Why? The proprietary hub driving them had been killed without a replacement or migration path, so all devices had to be replaced. That said, all of Google's major competitors are guilty of the same, and many smaller organizations are as well. Naturally, I am not suggesting that commercial organizations maintain unprofitable products forever. But why not open-source the code of those discontinued products? Wasting the years of development effort that went into them is waste humanity cannot sustain. It is shortsighted to think otherwise. In due time, regulators may force organizations to do the right thing. But how nice it would be if they could take the initiative for a change! By making your product (or at least its core) open source from the very start, you provide a way forward for your users if you wish to move on in the future while preserving your capacity to build meaningful commercial innovations. Remember that the true innovators of this century will be the organizations that not only use open source in their products and solutions but will acquire the skills required to contribute and give back to the community.

My goal in this book was to provide an overview of the Eclipse IoT ecosystem and give you the tools and knowledge you need to build enterprise-grade solutions. I may be naive, but I still sincerely believe in the potential for technology to improve our lives. All you need to do for that to happen is to take security seriously, defend user privacy, and take a long-term perspective.

And now, get your project started!

Index

A

Access Control Lists (ACLs), 31, 72, 107

Acknowledgments, 27, 28, 32, 69, 75, 313

Actuators, 5, 12–14, 16, 22, 164, 190–192, 194, 202, 204, 206, 212, 347

Adoptium working group, 19

Amazon Web Services (AWS), 250, 254, 287, 290

Analog-to-digital converters (ADCs), 199, 204, 216

Analytics, 8, 16, 17, 164, 201, 341, 342

Apache License v2.0, 225, 246

Apache Maven, 34, 96, 117

Apollo guidance computer, 67

Apollo Moon Missions, 7

Application Programming Interfaces (APIs), 12–15, 17–19, 61, 96–98, 122, 165–167, 175–177, 244, 252–254, 310–313

Application runtimes
  Jakarta EE, 299, 300
  Java EE, 299, 300
  .NET Core, 298, 299
  Node.js, 301
  Pythod, 302, 303
  Wasm, 304

Arduino, 11, 101, 198–199, 204, 215, 224, 258
  boards, 199, 258
  sketches, 258

Arm Cortex-A ecosystem
  big.LITTLE configuration, 211
  characteristics
    64-bit, 208–210
    32-bit, 207, 208
  chipmakers, 211
  cores, 207
  license, 211
  processor designers, 211
  variants, 211

Arm Keil Studio, 249

Arm Mbed OS, 246–250

Auto-discovery mechanisms, 128, 162, 178

B

Bare metal, 243, 248
  Arduino boards, 258
  Drogue IoT, 259, 260
  ESP-IDF, 261, 262

Basic Rate/Enhanced Data Rate (BR/EDR), 220

Birth certificate, 108, 111, 112

Bluetooth
  advertising interval, 220
  Bluetooth LE, 220
  BR/EDR, 220
  callback functions, 240–242
  declaration, 241
  frequency band, 219
  frequency-hopping spread spectrum, 220
  Mesh networking, 239
  milliwatts/range, 219
  models, 239, 240

© Frédéric Desbiens 2023
F. Desbiens, *Building Enterprise IoT Solutions with Eclipse IoT Technologies*,
https://doi.org/10.1007/978-1-4842-8882-5

Bluetooth (*cont.*)
 operations, 241
 protocols, 242
 provisioning, 239, 240
 researchers, 221
 security, 221
 6LoWPAN, 221
 structs, 240
 types, 220
 Zephyr, 238, 239
Bluetooth Low Energy (BLE), 30, 202, 220,
 221, 229, 239, 249, 255, 260
Bootstrap from Smartcard, 49
Bootstrap interface, 49
British Broadcasting Corporation (BBC), 215

C

CBOR Object Signing and Encryption
 (COSE), 32
Cellular Internet of Things (CIoT), 51, 53
Certificate mode, 31
cf-helloworld-client code, 35, 36, 38
cf-helloworld-client demo, 34
cleanSession, 78, 108
clientId, 78
Client initiated bootstrap, 49
Client registration interface, 49
Cloud, 6, 8, 14, 19, 20, 121, 192, 218, 233,
 261, 271, 273, 280, 290
Cloud-based servers, 10
Cloud Native applications, 272, 275, 292
Cloud2Edge, 288, 315
CoapResource class, 39
CoapResponse, 36
CoapServer class, 38
Complex data models, 8
Computerization, 3, 325

Concise Binary Object Representation
 (CBOR), 32, 48, 51, 52
CONNECT packets, 78, 81, 85, 108, 112
Connectivity, 4, 10, 217
 factors, 218, 219
 TCP/IP, 218
ConnectivityStatistics, 59
Consistent system, 14
Constrained Application Protocol (CoAP)
 acknowledgment message, 28
 Californium (*see* Eclipse Californium)
 characteristics, 26–29
 constrained devices, 42, 43
 DTLS protocol version 1.2, 26
 implementation, 42–44
 IoT protocols, 28
 lightweight reliability
  mechanism, 27
 message types, 27–28
 nonconfirmable messages, 28
 protocol stack, 29, 30
 reliability behavior, message
  types, 27, 28
 requests, 26, 27
 request URI, 28
 resources, 29
 RFC, 26
 security measures, 31, 32
Constrained devices, 10, 42, 100
 API, 12
 embedded developers, 12
 open source options, 12
 OS/RTOS, 12
 sensors and actuators, 13
 32-bit processors, 12
 vulnerabilities, 13
 zero-trust approach, 13
Constructor, 39, 98, 100

Continuous integration and
  delivery (CI/CD), 19, 272
"Continuous Session Awareness", 105, 107
Controller Area Network (CAN), 79, 201,
  207, 260
CORE-V IDE, 11
CredentialsUtil class, 42
Cyclonedds-python, 152, 153

D

DASH7 Alliance Protocol (D7A), 225
Data, 341
  Eclipse Streamsheets, 342–345
  IoT deployments, 341
  IoT devices, 341
Data-Centric Publish-Subscribe (DCPS),
  122, 125, 126
Data Distribution Service (DDS), 155, 158,
  162, 165, 173
  data-centric model, 129
  data writing, 139
  DDSI-RTPS, 124
  development, 122
  Eclipse Cyclone DDS (see Eclipse
    Cyclone DDS)
  GDS, 128
  implementations, 123
  IoT protocols, 122
  MQTT, 122
  protocol stack, 123, 124
  publishers and subscribers, 125–127
  QoS
    configuration, 134
    data availability, 130, 131
    data delivery, 131, 132
    data timeliness, 133
    resources, 133
    reading, 140, 141
    security, 141, 142
    specifications, 122, 123, 143
    topic
      filtering, 138, 139
      topic keys, instances and samples,
        137, 139
      topic types, 122, 134–136
Datagram Transport Layer Security
  (DTLS), 26, 30–33, 40, 49, 51–53,
  63, 66, 142, 232–235, 248
Data integration, 17, 325
Data modeling, 17, 325
Death certificate, 108, 111, 112, 114
Delivery guarantees, 73, 74, 132
Development tools, 10, 11, 15, 248, 249,
  251, 257, 259, 298
Device connectivity platforms, 290, 298,
  313, 326
Device LwM2M object, 47
Device management
  Eclipse hawkBit, 334–338
  Eclipse Kapua, 338–340
Device Management and Service
  Enablement Interface, 48, 50
Device Management Federation (DMF),
  18, 336, 338
DevOps, 7, 14, 18–20, 272, 278, 279
Digital twins, 19, 254, 285, 298,
  310–313, 317–319
Direct over-the-air (OTA)
  updates, 279
Distributed computing, 6, 14, 217, 272,
  273, 276
Ditto, 313
  architecture, 314
  AuthenticationProvider, 316
  Client SDK, 315

Ditto (*cont.*)

    as a Cloud back end, 311

    connections, 313

    digital twin platform, 311

    Eclipse Foundation, 311

    entities, 312

    functional scope, 311

    installation instructions, 315

    JSON payloads, 313

    MessagingProvider, 316

    microservices, 314

    nginx.conf file, 315

    official web resources, 311

    repository, 317

    REST API, 318

    as a WoT TD, 318

Docker deployment, 315

Domain Name System (DNS), 78, 95, 234, 252

Domain-specific proprietary field buses, 79

Downstream messages, 327

Drogue Device, 260, 261

Drogue IoT, 243, 259–261, 303

DTLSConnector, 40, 41

E

Eclipse Arrowhead, 20

Eclipse Californium

    Apache Maven, 34

    cf-helloworld-client code, 35, 36

    cf-helloworld-client demo, 34

    CoAP server code, 38, 39

    CoAP server demo, 37, 38

    DTLS, 41, 42

    features, 33

    official web resources, 32

    repository, 32

    sandbox server, 33

Eclipse CDT, 11, 251, 257

Eclipse Che, 11

Eclipse community, 9, 317

Eclipse Cyclone DDS, 121, 143

    Cyclone DDS, 143, 153, 154

    DDS specifications, 143

    "Hello World" application, 147

        data model, 147, 148

        publishing, 148–151

        python version, 152, 153

        running sample, 151, 152

    installation

        cloning the repository, 144

        code compilation, 144, 145

        software packages, 143

        testing your setup, 146

Eclipse Dataspace Connector project, 17

Eclipse Development Process, 22

Eclipse Distribution License v1.0, 32, 53, 88, 96, 143

Eclipse Ditto project, 19

Eclipse EdiTDor, 17, 306–307, 317

Eclipse Embedded CDT, 11, 251

Eclipse fog05, 15

Eclipse Foundation, 9–11, 18, 21–23, 82, 103, 143, 182, 194, 225, 257, 274, 280, 285, 300, 311, 321

Eclipse Foundation protocol implementations, 21

Eclipse Foundation's IoT and Edge ecosystem, 22

Eclipse Foundation Specification Process (EFSP), 22, 104, 117, 305

Eclipse 4diac, 15, 297, 319–321

Eclipse Hara project, 13

Eclipse hawkBit, 13, 18, 285, 333–337

Eclipse Hono, 18, 72, 285, 287, 288, 298, 313, 315, 326–328, 334, 341

Eclipse ioFog, 15, 282–285

Eclipse IoT
  ecosystem, 326, 350
  projects, 15, 32, 282, 288, 298, 315, 328
  working group, 5, 11, 15

Eclipse Kapua, 17, 288, 290, 326, 333, 338–340

Eclipse Keyple, 15, 297, 321–322

Eclipse Kura, 15, 17, 19, 280, 288–292, 338

Eclipse Leshan project
  client, 58–60
  Java libraries, 53
  main versions, 53
  official web resources, 53
  Sandbox Server, 54
  server, 61–63
  test servers, 54–57

Eclipse MRAA project, 12

Eclipse Paho, 82, 96, 100, 101
  Java version, 96–98
  Python version, 98–100

Eclipse Sparkplug, 103, 114
  See also Sparkplug

Eclipse Streamsheets, 17, 342–345

Eclipse Tahu, 104, 112, 117–119

Eclipse Temurin, 19, 299

Eclipse Theia, 11, 250, 251

Eclipse Thingweb, 5, 17, 306, 308–310

Eclipse UPM interfaces, 12

Eclipse VOLTTRON, 15, 297, 322–323

Eclipse Wakaama, 54

ECMAScript, 300, 306

Edge analytics, 16

Edge applications, 15, 294, 297, 298, 321

Edge Compute Network (ECN), 281–283, 285

Edge computing, 244, 262, 276, 280
  Cloud Native and DevOps approaches, 7
  definitions, 6
  distributed computing, 14, 272
  Eclipse ioFog, 281
    applications, 282
    ioFog concepts and architecture, 281–283
    on macOS, 283
    official web resources for ioFog, 281
  Eclipse Kanto, 285
    concepts and architecture, 285–287
    official web resources for Kanto, 285
  Eclipse Kura, 288
    concepts and architecture, 288–290
    Docker, 291
    graphical interface, 291
    official web resources for Kura, 288
  edge-to-cloud continuum, 274, 275
  Edge *vs.* Cloud, 273
  IoT implementation advantages, 7–9
  K3s, 14
  Kubernetes, 14, 280, 292–294
  microservices, 14
  projects, 276, 294–296

Edge environment, 7, 18, 273, 293

Edge gateways, 14, 15

Edge Native applications, 272, 274, 275, 277–279, 288

Edge nodes, 4, 10, 14, 20, 106–111, 115, 206, 262–263, 273, 276, 297

EdgeOps, 7, 18–20, 278–279

Edge orchestration, 14, 15

Edge servers, 14–16, 262

Edge-to-cloud continuum, 15, 273–275, 319

Edge Virtualization Engine (EVE), 294–296

Electrification, 3

Embassy, 260
Embedded systems, 67, 68, 262, 336
Ephemeral Diffie-Hellman (DHE), 233
Error correction code (ECC), 197, 211
Espressif IoT Development Framework
(ESP-IDF), 261–262
Ethernet, 10, 30, 190, 201, 211, 219, 223,
224, 228, 232, 236
Executables, 15, 47, 146, 175

F

Factory Bootstrap, 49
Field-by-field basis, 32
FreeRTOS, 249
AWS, 250, 254
hardware support, 250
kernel, 251, 252
libraries, 252–254
lowdown, 254
ports, 250, 251
Frequency-shift keying (FSK), 229

G

GAIA-X European project's protocols, 17
Gateways, 14–17, 52, 81, 104, 123, 200,
206–207, 218, 222–224, 235, 262,
280, 288, 290, 338
General-purpose input/output (GPIO),
204, 211, 215, 216, 261, 288
GitHub repository, 34, 98, 101, 152, 261,
287, 295, 315
GLaDOS, 115
Global Data Space (GDS), 127, 128
data-centric model, DDS, 129
domains, 128
DSS partitions, 129

Graphics processing units (GPUs),
194, 211
GSM Association (GSMA), 227

H

Hardware, 10
cost, 193, 194
environmental conditions, 189
life cycle, 190, 191
power consumption, 190
security, 192, 193
use case requirements, 191, 192
Hardware abstraction layers (HAL), 12,
191, 248, 259, 260
Hardware support, 218, 244, 250, 257, 260,
262, 273
hawkBit
access, 337
characteristics, APIs, 336
DDI and DMF APIs, 338
device and software repository, 334
documentation, 336
features, 334
microservices, 334
official web resources, 334
rollout and campaign
management, 335
software module, 334
software updates management, 334
user interface, 335
HelloWorldResource, 39
Historian databases, 8, 71, 167, 323, 342
Hono
AMQP 1.0, 328
architecture, 326
authentication, 331
CLI, 329, 333

components, 328

device registry, 327

Eclipse Public License
    version 2.0, 326

messaging APIs, 327

MQTT protocol adapter, 332

multitenant platform, 327, 330

official web resources, 326

platform, 328

sandbox, 329

I

Industrial Internet of Things (IIoT),
    6, 104–106

Industry 4.0, 6, 280

Information technology (IT), 279, 325, 341

Integrated circuit (IC), 191, 194, 204, 205,
    222, 247

Integrated Development Environment
    (IDE), 11, 249, 251, 257, 258,
    262, 319–321

Inter-integrated Circuit (I²C), 204–206,
    211, 214–216, 248, 259–261, 288

International Data Spaces standard
    (IDS), 17

International mobile subscriber identity
    (IMSI), 222

Internet, 3, 4, 7, 8, 11, 12, 20, 67, 121, 155,
    218, 222, 260, 271, 279

Internet Engineering Task Force (IETF),
    21, 26, 249

Internet of Things (IoT)

    Bluetooth, 3

    concepts, 5

    definitions, 4

    developers, 212

    devices, 5, 6, 67

edge computing, 6–9, 206, 218

edge protocol implementation
    projects, 21

    and edge toolkit, 22, 23

endpoints, 206

gateways, 207

platforms

    AdoptOpenJDK initiative, 19

    analytics, 16

    API management, 19

    applications, 18

    data integration, 17

    data management, 17

    device management, 18

    device registry and record, 18

    Eclipse project, 20

    Eclipse Streamsheets, 17

    EdgeOps, 19

    Hono supports, 18

    infrastructure, 16

    languages, 18

    ML, 16

    networks and protocols, 18

    related concepts, 16

protocols, 20, 333

reference architecture, 9

    common layers, 10, 11

    constrained devices, 12, 13

    definition, 9

    Eclipse community, 9

    edges, 14–16

    IoT platforms, 16–20

    project logos, 23

sensors, 5

specific protocols, 20

Wi-Fi networks, 3

ioFog architecture, 282

ioFog Controller, 282

J

Jakarta EE, 19, 104, 299–300, 326
Java Development Kit (JDK), 42, 54
Java Enterprise Edition (Java EE), 18, 104,
        299, 300

K

Kapua, 338
    application integration, 339
    data management, 339
    device connectivity, 338
    device management, 338
    features, 338
    Linux/macOS, 340
    official web resources, 338
    security model, 339
    web user interface, 339
K3s packages, 293
KubeEdge, 14, 280, 292–294
Kubernetes, 14, 15, 279–281, 292–294, 314,
        315, 328
Kura documentation, 292
Kura's Cloud APIs, 290

L

Last Will and Testament (LWT) feature,
        77, 78, 81, 86, 107, 108
Latency, 7–8, 16, 22, 73, 74, 105, 121, 146,
        179, 191, 217, 218, 225, 227, 229,
        235, 245, 325
LeshanClientBuilder class, 59
Leshan Demo Device, 56
Leshan demo server, 54, 57
LeshanServerDemo class, 63
Leshan wiki, 53
Libmosquitto, 96

Lightweight Machine-to-Machine
        (LwM2M), 18
    aims, 46
    CoAP, 45
        bootstrap, 49
        client registration, 49
        device management/service
            enablement, 50
        information reporting, 50
    constrained devices, 63–66
    interfaces, 48–50
    IPSP node, 65
    lwm2m_setup function, 64
    objects, 46, 47
    OMA SpecWorks, 46
    protocol stack, 52, 53
    registry, 46, 48
    resources, 46, 47
    Sensor Value (ID 5700), 46
    6LoWPAN, 65–66
    specification, 47, 48
    third parties, 46
    versions
        LwM2M v1.0, 50, 51
        LwM2M v1.1, 51
        LwM2M v1.2, 51
Linux, 12–15, 30, 42, 54, 65, 88, 101, 143,
        154, 171, 255, 262–263, 265–267,
        301, 315, 336, 340
Local area networks
        (LAN), 7–8, 165, 223
Log4j project, 348
Long Range wide-area network
        (LoRaWAN), 7, 51, 53, 202, 225,
        226, 249, 260
Long-Term Evolution (LTE), 221–223, 225,
        227, 232, 235
Long-Term Support (LTS), 251, 257

Low-power wide-area networks
(LPWANs), 224, 225, 227
LTE-M, 202, 222, 225, 227, 235

M

Machine learning (ML), 8, 16, 341, 347
Machine to machine (M2M)
communication, 5
Manufacturer resource, 48, 56
Mbed CLI 2, 249
Mbed OS APIs, 248–249
Mbed OS' architecture, 246, 247
Mbed Studio, 249, 250
Mean time between failures (MTBF), 207
Memory management, 13, 100, 155, 245,
251, 252, 262
Memory management unit (MMU),
245, 262
Memory protection hardware (MPU), 245
Metropolitan area networks (MAN), 223
Micro:bit, 215, 239, 261
Microcontrollers (MCUs), 67, 194, 297
Arm architecture, 195
core features, 196, 197
Cortex-M, 195, 196
Cortex-M4, 195
Cortex-M7, 197
Cortex-R, 195
AVR architecture, 198, 199
choices, 201, 202
CORE-V, 214
designs, 194
Harvard architecture, 195
I/O interfaces, 205
I/O protocols, 206
RISC-V architecture, 199, 200
von Neumann architecture, 194

Microprofile project, 18
Microservice-based distributed
applications, 277
Modern edge computing, 14
Mosquitto
configuration, 89–91
Eclipse Distribution License v1.0, 88
Eclipse Public License v2.0, 88
installation, 88, 89
official web resources, 88
sandbox servers, 95
setting up TLS, 92–95
2.0.14, 88
MQ Telemetry Transport (MQTT), 67
clients, 68
connections and sessions, 76–78
and constrained devices, 100
crucial design decision, 69
Edge Nodes, 106
growth and adoption, 103
invention, 68
messages, 68
fixed header, 69
payload, 69
variable header, 69
and MQTT-SN, 81, 82
OASIS Open standard, 68
protocol stack, 79, 80
publishing and subscribing, 74–76
publishers and subscribers, 68
QoS, 73, 74
security, 87, 88
Server (broker), 106
specification, 69, 103
subscription return codes, 76
topic filters, 70, 71
topics, 70, 71
version 5.0, 68, 82 (*See also* MQTT v5.0)

MQTTPacket, 101

MQTT-SN, 80, 81

   clients, 81

   gateways, 81

   and MQTT, 81, 82

MQTTSNPacket, 101

MQTT v3.1.1, 107

MQTT v5.0, 82, 107

   AUTH packets, 85

   broker reference, 87

   Clean Session flag, 84

   DISCONNECT packet, 85

   flow control, 86

   LWT delay, 86

   message expiration, 84

   optional features, 84

   passwords without
      usernames, 85

   reason codes, 83

   request/response interactions, 84

   shared subscriptions, 85, 86

   user-defined properties, 83

Multiple processor architectures, 297

N

Narrowband

   DASH7 protocol, 225

   Internet of Things, 227

   LoRaWAN, 226

   LPWANs, 225

   options, 7, 225

National Electrical Manufacturers
      Association (NEMA), 193

.NET Core, 298–299

Network, 3–5, 7, 8, 12, 18, 30, 52, 80, 105,
      123, 217, 220, 221, 230–232, 277,
      325, 334

Network Address Translation (NAT)
      proxy, 30, 51, 52, 76

Network design considerations

   confidentiality, 232–234

   out-of-band management, 236

   protocols, 232

   redundant connections, 235, 236

   resiliency, 234, 235

Nginx, 296, 314, 315

Node.js runtime
      environment, 301

Non-Eclipse DDI client libraries, 336

NoSec, 31, 32

O

ObjectsInitializer class, 60

OpenHW Group, 10, 11, 191, 213, 214

Open source

   dependency, 25

   hardware, 212

   projects and standards, 25

   RISC-V architecture, 10

   shared functionalities, 25

OpenSSL, 144, 348

Operating system (OS), 12, 243

   APIs, 244

   hardware support, 244

   mainframes, 243

   memory management, 245

   multitasking, 245

   real-time requirements, 245

   RTOSes (*see* Real-Time Operating
      Systems (RTOSes))

   shared services, 245

Operational technology (OT), 5, 279,
      325, 341

OSCORE extension (RFC 8613), 32

P, Q

Payload Format, 69, 105, 106

PCI Express (PCIe), 211, 222

Peripheral Access Crate (PAC), 259

Piggybacked acknowledgments, 28

Power-efficient processor architectures, 10

Power over Ethernet (PoE), 190, 223, 224

PreSharedKey, 31

Privacy, 16, 221, 306, 347–350

Private Cloud, 271

Processor cores
    CVA6 Family, 213
    CVE4 Family, 213, 214
    OpenHW CORE-V Family, 213

Programmable logic controllers (PLCs), 14, 106, 207, 297, 320, 321

Protocol buffers, 112

Protocols, 11, 15, 17, 18, 20–22, 25–28, 107, 134, 155, 190, 225, 232, 235, 295, 305, 325, 342

PUBACK packet, 74

Public Cloud, 271–273

Public Key Infrastructure (PKI), 31, 51

PUBLISH packet, 69, 74, 75, 79

Publish/subscribe model, 20

Pulse-width modulation (PWM), 206, 214–216

Python, 98–100, 302–303
    binding, 173
    environment, 264, 266
    versions, 152–153

Python Package Index (PyPI), 98, 153, 302

R

Radio-frequency identification (RFID), 225

Radio Resource Control (RRC), 227

Raspberry Pi, 12, 89, 93, 192, 216, 224, 257, 291, 344

RawPublicKey, 31, 32

Real-Time Operating Systems (RTOSes), 12
    Arm Mbed OS, 246–250
    FreeRTOS (see FreeRTOS)
    Zephyr, 255–257

Real-Time Publish-Subscribe Protocol (RTPS), 122, 124

Real-time systems, 245

RegistrationService interface, 61

Regulated industries, 8

Request for Comments (RFC), 26–29, 31–33, 42, 170, 221, 253, 274, 308, 334

Request/response model, 20

Reset, 28

Resource Directory (RD) client, 63

Retained messages, 74, 78–79, 84, 90, 107

RISC-V ecosystem, 213

Robotics, 153–154, 206

Rust, 96, 155, 165, 171, 172, 259–261, 303–306, 336

S

Sandbox server, 33, 54, 95

Second generation (2G), 3

Security, 10, 31–32, 87–88, 141–143, 192–193, 290, 347

send_simple_coap_request function, 43

Sensors and actuators, 16, 347
    characteristics, 202, 203
    connection
        ADCs, 204
        GPIO, 204
        I^{2}C, 205

Sensors and actuators (*cont.*)
    IC, 204
    SPI, 205
    UART, 205
    1-Wire, 205
  MCU I/O protocols, 206
  uses, 204
Sensor Value, 46, 48
Serial ATA (SATA), 211
Serialization formats, 48, 51, 52, 63
Serial Peripheral Interface (SPI), 205, 211,
    214, 215
Server initiated bootstrap, 49
Service Plugin Interface (SPI), 123, 141
Shared subscriptions, 85–86
Short Message Service (SMS), 51, 52
Short-sightedness, 347, 349, 350
Simple Network Time Protocol
    (SNTP), 253
Single-board computers
  Arduino, 215
  micro:bit, 215
  Raspberry Pi, 216
Small form-factor pluggable (SFP), 223
Software-as-a-Service (SaaS), 84, 107, 279
Software-Defined Networking (SDN), 274
Software development kits (SDKs), 202,
    257, 259, 263, 265–267, 311,
    313, 315–317
Sparkplug, 22, 68
  architecture, 106
  components, 106
  connector, 104
  design principles, 107, 108
  devices, 115
  edge nodes, 115
  evolution, 104
  fundamental rule, 108

  on GitHub, 104
  Host Application, 106, 115
  and interoperability, 104, 105
  messages types
    DBIRTH, 111
    DCMD, 112
    DDATA, 111
    DDEATH, 111
    NBIRTH, 111
    NCMD, 111
    NDATA, 111
    NDEATH, 111
    STATE, 112
  metrics, 113
  MQTT requirements, 107
  payload definition, 112
  payloads, 105, 113
  session management, 114
  specification, 104, 112
  state management, 105
  topic namespace, 105, 109
    device ID, 110
    edge node IDs, 110
    group ID element, 109
    message types, 110
    namespace element, 109
  uuid, 113
start_coap_client function, 43
Storage/data encryption, 192
Streamsheets
  commands, 344
  comprehensive security, 343
  container images, 344
  data flows, 343
  data streams, 342
  and Mosquitto 2.0, 344
  official web resources, 342
  user interface, 343

SUBSCRIBE packet, 75, 76

Supervisory Control and Data Acquisition (SCADA), 67, 321

Swiss knives, 20

Systems on a chip (SOCs), 191, 194, 199, 200, 262

T

TCP/IP-based alternatives, 79

Thing Models (TM), 306, 307

3rd Generation Partnership Project (3GPP), 53, 221, 227

TLS/DTLS, 49, 232–235

Topic filters, 70–72, 75

Topic names, 70–72, 76, 81, 104–106, 109, 110, 114

Traditional wired (Ethernet), 30

Transport Layer Security (TLS), 26, 31, 52, 53, 78, 80, 87, 90, 92–95, 142, 144, 170, 232, 234, 248, 343

True Random Number Generator (TRNG), 247

Trusted-Firmware Cortex-M technologies, 247

Trusted Platform Module (TPM), 192, 254

U

Uniform Resource Locators (URLs), 26, 57, 165, 172, 285, 309, 316, 318, 329, 345

Unikernels, 244

Univac 1230 mainframes, 6

Universal Asynchronous Receiver-Transmitters (UARTs), 201, 205, 211, 214, 216, 236

Universal Serial Bus (USB), 192, 201, 211, 222, 260, 288

Upstream messages, 327

User Datagram Protocol (UDP), 20, 26, 27, 31, 38, 51, 52, 80, 123, 142, 170, 232, 319

UTF-8, 69–71, 83, 109, 110, 112, 158, 178, 181

V

Value resources, 47

Variable byte integer, 69

Vendor-neutral fashion, 22

Voice over Long-Term Evolution (VoLTE), 222, 227

W

WebAssembly (Wasm), 304, 305

Web of Things (WoT), 5, 17, 298, 305, 306
  Eclipse EdiTDor, 306, 307
  Eclipse Thingweb, 308
  EdiTDor team, 307
  Node.js, 308

Wide area networks (WAN), 165, 223, 224

Wi-Fi
  Alliance, 228
  constrained device, 238
  definition, 228
  ESP32 Wi-Fi driver, 238
  main method, 238
  modules, 237
  operational range, 229
  radio bands, 228
  scan command, 237
  versions, 228
  Wi-Fi HaLow, 229
  Zephyr, 237

Windows, 14, 42, 88, 143, 144, 154, 171, 172, 182, 257, 266–267, 297–299, 302, 340, 345

Wireless Personal Area Networks
(WPAN), 30
Wireless (Wi-Fi) networks, 30
World Wide Web Consortium (W3C), 5,
17, 298, 304, 305

X, Y

x86-64 ecosystem, 211, 212

Z

Zenoh protocol
back ends, 166
constrained devices, 181
installation, 182
publishing, 184, 185
subscribing, 183, 184
Zephyr, 182, 183
core implementation, 156
data life cycle, 156
data retrieval, 156
deployment units
nodes, 160–162
topologies, 160, 161
device discovery, 162
installation
C binding, 172
core library, 171, 172
Python binding, 173
router, 175–178
Rust, 171
testing, 173
key abstractions
distributed queries, 158
encodings benefits, 158, 159
expressions, 157
keys, 157

MQTT, 158
resources, 157
router-less topologies, 159
selectors, 157, 158
location-transparent primitives, 155
MQTT, 156
plain subscriptions, 156
plugins
DDS, 165
REST, 165
storages, 165–167
webserver, 165
primitives, 159
protocol stack, 170
publishing, 180, 181
reliability and congestion control
hop-to-hop, 167
layers, 167
publishers control, 168
strategies, 167, 168
subscribers control, 168
robots, 163, 164
scope, 156
scouting, 178, 179
subscription, 179, 180
timestamps, 159
web resources, 171
zenoh-flow, 168, 169
Zephyr, 255–257, 263–266
LwM2M, 65
operating system, 243
repository, 42
RTOS, 42, 63, 100, 101
Zero Touch onboarding, 278
Zero-trust approach, 10, 13
Zero Trust security model, 278
Zigbee, 7, 10, 30, 80, 230–231
Z-Wave, 10, 229–232

Printed in the United States
by Baker & Taylor Publisher Services